Positive Psychology

Applications in Work, Health and Well-being

Edited by

Updesh Kumar

Archana

VIjay Parkash

Pearson

Acquisitions Editor: Kaushal Jajware
Production Editor: Vipin Kumar

ISBN 978-93-325-4131-3

First Impression

Published by Pearson India Education Services Pvt. Ltd, CIN: U72200TN2005PTC057128, formerly known as TutorVista Global Pvt. Ltd, licensee of Pearson Education in South Asia.

Head Office: A-8(A), 7th Floor, Knowledge Boulevard, Sector 62, Noida 201 309, Uttar Pradesh, India.
Registered Office: Module G4, Ground Floor, Elnet Software City, TS-140, Block 2 & 9, Rajiv Gandhi Salai, Taramani, Chennai 600 113, Tamil Nadu, India.
Fax: 080-30461003, Phone: 080-30461060
www.pearson.co.in, Email: companysecretary.india@pearson.com

Typeset by *Content Management Team India*
Printed in India at Repro India Limited

Contents

Preface

There are numerous disciplines and sub-disciplines in the field of science and many schools of thought exist in different social science disciplines where the concepts are relatively more abstract than concrete in nature. Positive psychology is one such sub-discipline within the realm of psychological sciences. However, rather than considering it a separate branch of psychology, we believe that it is a recently emerging school of thought that has gained massive momentum over the past few years. In general, psychology—with human behaviour as its focus of attention—relentlessly strives for and centers around welfare of human beings across various domains of behavioural operations, and therefore, we fail to recognize any negative side of psychology. Even if the subject matter of study for different branches or sub-branches of psychology covers certain maladaptive behaviour patterns or psychological illnesses, the ultimate goal before a psychologist is work towards the betterment of humankind in order to cultivate a positive and growing world. It is considered that positive psychology has emerged as one of the recent branches of psychology and it has wider range of applications in the area of education, health, stress management, self-help skills and work settings. This school of thought gives impetus on the perspective of building efficiency to maintain positive and well-adapted life style rather than identifying deficiencies and providing interventions to restore adaptive mechanisms. It focuses on building good lives and not only on the actualization of the individual but on the development of other people and the community within the framework of their contributions.

Positive psychology is an approach that has rapidly grown during the past decade and contineus to capture attention of many researchers working in the area of clinical, counseling and health psychology. This growing interest of the researchers has led to enormous addition to the literature and research findings in this field. However, most of the literary efforts devoted so far by some of the eminent scholars in the field primarily concentrate on elaborating the conceptual arena and revolve around comprehension and description of the constructs of positive psychology. There is a relative paucity of explanations of application of guiding principles of this emerging branch of psychology in varied practical realms of life. Further, the existing books in this area include contributions from a few numbers of researchers and scholars in this area and mainly represent Western ideology. Therefore, we believe that there is a need of bringing forth a volume that primarily discusses the implications and applications of this school of thought rather than simply describing and revisiting the conceptual ideology. *Positive Psychology: Applications in Work, Health and Well-being* intends to elaborate upon the applications of positive psychology in work, health and well-being to bring about desired positive outcomes. This volume is an attempt to provide a wider perspective to the reader by providing a comprehensive synthesis of Western and Eastern perspectives. It is an endeavour to extend the journey of positive psychology by further enhancing the conceptual awareness from the perspectives of Indian scholars and discussing the ways in which the approaches of this field serve as stress-busters to deal with life challenges, help people lead a qualitatively better life and create flourishing organizations. The volume incorporates fourteen chapters authored by eminent academicians and scholars from reputed academic institutions across India. The chapters have been clubbed under two broad sections. The first section on *Theoretical Perspectives* deals with the comprehensive understanding of general issues of positive psychology and the second section discusses its *Applications in Work, Health and Well-being.*

As Professor K. Ramakrishna Rao states in his foreword that there seems to be a link and affinity between Indian psychology and positive psychology, the volume opens with delineating of Indian perspectives on positive psychology. In the first chapter *Indian Perspectives and Positive Psychology,* the author Kiran Kumar K. Salagame, an eminent academician and researcher in this field, discusses about the development of the field of positive psychology in Western world and brings out the bridging similarities between Indian traditional dogmas and the prevalent positive psychology conceptualizations. He puts forth a very relevant discourse about different underlying paradigms to understand psychological insights prevalent across the world. He puts across Indian perspectives on positive psychology by elaborating on *Vedic* and *non-Vedic* Indian ideological and philosophical understandings. Extending his discourse further in the second chapter of the volume, Jitendra K Singh, a senior psychology scientist at the Defence Institute of Psychological Research, concisely delineates upon the *Progress in Positive Psychology* and in doing so he also presents some reflections from India. Pointing to the contemporary developments in the field of positive psychology, the author cites some recent empirical works and publications. He further puts forth some empirically established Indian conceptualizations for the existing western connotations and establishes the validity of his argument by citing empirical researches from Indian pioneers.

The field of positive psychology—as it is widely known across the academic world—primarily builds around the concepts of virtues and strengths within the individual. This volume takes shape with the third chapter on *Virtues and Well-being*, wherein Surendra Kumar Sia very nicely elaborates the concept of virtues from varied viewpoints like Buddhism, Confucianism, and Hinduism. He further provides an empirical synthesis on the concept and highlights the applications of virtues in promoting well-being and quality of life. Further research implications have also been well-documented by the author. Pioneers of this field of positive psychology have enumerated twenty-four character strengths under six virtues. The volume proceeds further in a sequential manner incorporating the fourth chapter on *Character Strengths and Health* authored by Amrita Yadava. She opens the discussion by attempting to explain the health–behaviour linkage in a very logical and empirically proven way. Putting forth a comprehensive note on character strengths, she takes the discourse to the applications of an individual's character strengths in maintaining his/her physical and mental health and establishes a strong and valid relationship among these.

Extending the scope of volume further, the fifth chapter of this volume authored by Hardeep Lal Joshi delineates on *Happiness and Quality of Life.* In a comprehensively synthesized form, the author elaborates upon the concepts of happiness and quality of life in light of different theoretical models and very concisely discusses about various positive psychological correlates of happiness and well-being by citing research-based instances. The concept of happiness has been taken forward by Akbar Husain and Shahin Zehra in the next chapter wherein they present *An Analytical Model of Happiness* before the reader. In this chapter, authors provide an elaboration on the psycho-social correlates of happiness and assertively opine that in a more practical way happiness can be defined as composite of Health, Attributional style, Pleasure, Positive emotions, Interpersonal relationships, Nature, Extraversion, Spirituality, and Self-esteem. By means of these acronyms, they explain a new theoretical model of happiness that explains the concept in a comprehensive manner thus leading to future researches in this domain.

The second section of the volume focuses more on the applied aspects of positive psychology and opens with the seventh chapter authored by Papri Nath and Rabindra Kumar Pradhan. The chapter extends the applications of positive psychology in organizational settings, and herein, the authors establish the *Psychological Capital as a Catalyst to Positive Organizations.* After providing conceptual outlines for positive organizations, positive organizational behaviour and psychological capital, they envisage various significant linkages of psychological capital with enhanced job performance, job commitment and job satisfaction, better health and well-being, and reduced job stress and absenteeism. The authors also elaborate upon how psychological capital helps improve employee engagement and job involvement along with creating a supportive and satisfying work climate. They further propose a conceptual framework wherein psychological capital mediates organizational climate and processes to provide positive outcomes. Moving forth from understanding the applications of positive psychology in traditional organizations and entering into the domain of positive interpersonal relationships, Rana and Hariharan very nicely talk about *Positive Social Support and Flourishing Relationships* in the eighth chapter of the volume. They adequately elucidate the conceptual perceptive of social support and its role as stress bufferer and a way to remain healthy. With the help of various theoretical models, they systematically elaborate about how social support can help in flourishing relationships and call for promotion and sustenance of qualitative social support for healthy psychological environment. Interpersonal relations play a significant part in all spheres of life and generally in all kinds of settings. However, family and workplace are the two spheres where an individual spends most of the time of his/her life and interpersonal interactions are a must at these places. With a related concern in mind and focusing the attention on another domain of interpersonal relationships, Chadha and Chopra call for *Positive Work–Family Interface* in the next chapter. Highlighting the aspects of conflicts and interference, they describe the concept of work–family interface in a concise manner drawing heavily on numerous existing empirical studies. Further into this discourse, they elaborate upon the positive side—enrichment perspective—of the interface between work and family domains. Putting forth the empirically cited positive outcomes of the positive work-family interface, the authors conclude the chapter by listing some strategies for enhancing positive work-family interface and see its roots in the realm of positive psychology.

Bringing into note another emerging concept of positive psychology, Sanyal, in the tenth chapter of this volume, highlights the role of *Mindfulness Meditation* as an *Essential Stress-buster in Military Life Context.* She attempts to discuss the stressors and stressful aspects of military life and puts forth the beneficial dynamism of mindfulness training which has a positive connectivity with enhanced well-being. Her special impetus is on utility of mindfulness meditation in military and accordingly, she has provided ample elucidations about the concept and ingredients of mindfulness meditation. Giving a call for utilising mindfulness-based stress reduction training for military personnel, she puts an emphasis on reaching to stress-free realization and actualization stages. Extending the discourse on applications of positive psychology conceptualizations to military settings in the next chapter, Rajbir Singh and Lokesh Gupta elaborate upon *Resilience and Its Relevance in Military Context.* They put forth thorough deliberations on the conceptual understandings of resilience at personal individual level and collective levels and provide a resilience framework for military settings. With adequate explanation of various risk and protective factors, the authors give an impetus on the need for building resilience among soldiers making a tough and integrated military environment.

Prevalence of and dealing with stress by means of positive psychology applications remains central to the twelfth chapter as well. In this chapter, Sibnath Deb and his colleagues discuss about *Stress and Resilience Capacity of Students* and consider this as *a burning public health issue.* With brief note on stress in general, they focus on stress among pre-school and school children and highlight various influencing and stress inducing factors prevalent at schools and home. In light of Indian beliefs and positive psychology principles, they talk of resilience capacity of students and call for various preventive programs for checking stress among students.

Proceeding towards the end of this volume, in the last but one chapter Nov Rattan Sharma presents deliberations about *Applied Positive Psychology* in order to *enable communities to thrive.* He comprehensively enumerates various applications of the field of positive psychology at a broader community level over and above the individualistic ones. He puts forth his views with evidence based support drawn out of his own and his colleagues' extensive researches in this area. The chapter concludes with appreciable likely positive contributions of applied psychology in all spheres of life and the wishes of creating environment of happiness for everyone. Lots of applications of positive psychology could be seen in stress, health, well-being, happiness, and quality of life, through the deliberations put across so far. Concluding the volume, we, the volume editors, have attempted in the last chapter to elucidate upon the applications of *Positive Psychology in Suicide Prevention.* With a brief conceptual note on suicidal behaviour, we have attempted to comprehend the prevalence of risk and protective factors and the interplay between the two. Further, on the preventive side, we have tried to devote our efforts to comprehensively describe the significant applications and protective roles of positive psychology principles like hope, social support, optimism, positive emotions and resilience in safeguarding the people against indulgence in suicidal behaviours. Since suicide is one of the leading causes of deaths across the globe, we call for more researches in this area based on positive psychology principles.

Although including everything about the whole positive psychology in one volume may not be a possible task, this school of thought is continuously growing with enormous pace. This volume might have left few aspects uncovered but a sincere effort has been made to reveal the applied side of positive psychology and most of the specifically relevant areas have been focused and thoroughly deliberated upon. As many texts exist in this field that put across the perspectives of western researchers, this volume brings forth the perspective of eminent Indian scholars actively engaged in researches in this arena. Although this may be taken as one critique of this volume, we believe that presenting the Indian perspectives before the reader will serve the fraternity to widen the horizon of culture specific conceptual understanding and future research latitude. Besides continuous duty-bound engagements, coming up with such a volume with amalgamation of diverse viewpoints from across an entire big country like India is not an easy affair and supportive channels come a long way in successful completion of the endeavour. In this context, we express our sincere gratitude to Professor Manas K Mandal, Distinguished Scientist and Director General (Life Sciences), Defence Research and Development Organization, Ministry of Defence for his inspirational instilling that kept us enthusiastically working upon this volume. We duly acknowledge all kind of unconditional support and encouragement provided to us during this venture by Dr K Ramachandran, the Director of our Institute, that is Defence Institute of Psychological Research. Lastly, we are very grateful to all the contributors for sharing their knowledge in the form of extensively worked out chapters to enlighten the readers and make qualitative additions to the existing literature in this domain.

The efforts put in the volume by us and by the revered authors are an attempt to extend the reach of positive psychology to larger masses, including students, researchers, academicians, social workers, psychiatrists and other health professionals. The volume has been designed to help readers stay healthy, just not mentally but in a holistic manner. It is a plan to bring about the applied aspects of positive psychology on a single platform by creating a bridge between Western and Eastern perspectives. We hope that the volume will provide significant positive inputs to the reader and will help ignite dedicated research efforts in this direction.

Updesh Kumar
Archana
Vijay Parkash

Foreword

Defence Institute of Psychological Research (DIPR), a division of the Defence Research and Development Organization (DRDO), Ministry of Defence, Government of India, is currently the most active centre for psychological research in the country. Its foundation was laid by W. Selvamurthy who is himself not a psychologist but a physiologist by training. However he has an abiding interest and deep knowledge of yoga and its relevance today. Following him, Manas Kumar Mandal has built the necessary structure for pursuing psychology from an Indian perspective.

At present, Updesh Kumar and his colleagues are continuing this tradition in a remarkable way. They have already made significant contributions in the field of psychology and have published some important titles. Kumar has edited five quality books on psychology; the present volume on positive psychology being one of them. This exemplifies that the work at the institute goes beyond the usual boundaries of defence to cover the latest trends in contemporary psychology. Positive psychology is clearly one of the front-runners with a promise of being at the centre of future progress in psychology.

Positive psychology has a short history but shows great promise for the future. It is essentially a 21st century phenomenon. Though the phrase positive psychology appeared in Abraham Maslow's book *Motivation and Personality*, first published in 1954, the launching of positive psychology as a system of psychology may be traced to the year 2000 when Martin Seligman and Mihaly Csikzentmihalyi published their paper 'Positive Psychology: An Introduction' in the *American Psychologist*. The rapid strides it made since then may be seen from the extraordinary number of publications.

In the XXXth International Congress of Psychology held in Cape Town, South Africa during 2012, I was amazed to find at least 80 papers of positive psychology being presented. There were 50 posters and eight seminars/workshops on this topic. What is equally amazing is that a great number of contributions came from China and India, far more than from USA and Europe. This made me wonder whether there are any 'archetypical' links between Asian psyche and positive psychology. This has led me to take a closer look at positive psychology from the perspective. The result is a target article 'Positive Psychology and Indian Psychology in Need of Mutual Reinforcement' (Rao, 2014a) published in *Psychological Studies*. This article was followed by several peer commentaries, including by some psychologists from other countries published in the same issue along with my response (Rao, 2014b).

All these developments are relevant to the context of publication of this volume, which on the one hand reinforces the expected link and affinity between Indian psychology and positive psychology and on the other hand it makes us cautious to avoid a trap. The trap is the temptation for uncritical imitation and replication of Western concepts and categories. The history of psychology in India as a university discipline is a good example how psychology as a discipline in India had suffered from such a syndrome.

The manifesto of positive psychology as proclaimed by Seligman and Csikszentmihalyi (2000) is 'to catalyze a change in psychology from a preoccupation only with repairing the worst things in life to also building the best qualities in life'. This is no different from the goal of Indian psychology which is transformation of and liberation from the existential limitations of the person to higher states of achievement, happiness and self-realization.

Indian psychology is not what many psychologists in India perceives it to be. As often noted, Indian psychologists by and large are generally engaged in imitating and replicating psychological research carried out in the West (Kao and Sinha, 1997). Indian psychology is not cultural psychology or cross-cultural psychology even though culture contributes significantly to the development of psychological thought. It is more than just indigenous psychology as it has global, pan-human relevance. It is not a branch or division of psychology like child psychology or industrial psychology. Indian psychology is a school/system of psychology like psychoanalysis. It is a system derived and constructed from classical Indian thought and psychospiritual practices prevalent in the Indian subcontinent for many, many centuries. Indian psychology offers model/s of human nature that stand in sharp contrast to the existing popular models in psychology. They have significant implications for future psychological research in India and beyond.

Indian psychology, in my model, is the study of the individual. The person is considered a composite of body, mind and consciousness. I call this the BMC Trident Model of the individual (Rao, 2011). Body is the physical structure that sustains the person. Consciousness is a fundamental principle of the universe in which the individual participates and derives his unique nature. Mind is the instrument that connects body and consciousness and makes it possible for setting up goals and achieve them in the human condition. The mind functions in three modes—cognition, emotion and volition. The three constitute the primary psychological processes in the individual. The goal of the individual is self-realization, which is none other than moving away from the moorings in the body to participate in consciousness per se and realize one's own true nature.

What is the link between Indian psychology (IP) and positive psychology (PP)? What are the points of convergence their perspectival differences? First, Indian psychology and positive psychology grew out of disenchantment with the current state of psychology. They are thus protest movements. Second both concepts focus on positive human strengths and how we can use them to promote happiness and consummate good. Third, both IP and PP claim to be culturally independent and have possible universal applicability.

Indian psychology has inherited a large mine of theoretical ideas and is currently left with the task of systematically gathering an appropriate empirical data base. Positive psychology has made a great beginning covering considerable empirical ground. However, it is weak in theory and, as I see it, there is need for conceptual clarity without which it is difficult to fit the data into a cohesive framework. In as much as IP and PP share the same disappointment with the current state of mainstream psychology and appear to take similar initiatives to correct the current situation, they may be seen in a complimentary relationship. Indian psychology can benefit by drawing on the data provided by positive psychology and the latter could gain much by embracing some of the theoretical constructs of Indian psychology. This can be illustrated by referring to two popular concepts in positive psychology—happiness and flow. Happiness, good life, fulfilling life are familiar terms embedded in the notion of subjective well-being in positive psychology and refer to different things and convey different meanings. For one thing, happiness may mean hedonistic pleasure and absence of pain. Indeed much of the research in positive psychology in this area is consonant with such meaning. The so-called philosophy of positive happiness beyond the hedonistic notions is little understood by positive psychologists. There is some recent research under the rubric of 'eudaimonia' to correct this and extend the meaning of happiness. As I see it, this attempt is piecemeal and insufficient to lead to systematic, programmatic research with a sound theoretical footing.

In this context, we have in Indian psychology rich theoretical support. Here we find a model of happiness more appropriately called 'bliss' (*ānanda*). Bliss is altruistic happiness which is distinct from hedonistic pleasure. Bliss is more than just mere absence of pain. It is a positive emotion, beyond sensuous pleasure inherent in human nature but something to be cultivated by practice. Hedonistic happiness is individual-centric. Altruistic happiness is transpersonal in that 'I' and the 'other' become reflexive of each other.

Ed Diener and associates, for example, refer to two distinct ways of looking at what constitutes happiness. One view is similar to what we have called altruistic happiness. According to this view, happiness consists in leading a fulfilled life characterized by love, wisdom and non-attachment. The other view is the utilitarian notion that considers absence of pain as the criterion of happiness. The utilitarians, the descendants of Jeremy Bentham, are 'the intellectual forerunners of subjective well-being researchers' (Diener, Lucas and Oishi, 2005, p. 63). The Indian model has ego-transcendence as central to happiness; therefore, altruism is the main characteristic of fulfilled life and happiness. In Gandhi's terminology, it is reducing oneself to zero, which is melting the ego into 'nothingness' and setting up goals of *sarvodaya* and non-possession. What is self-fulfillment in positive psychology is self-realization in Indian psychology.

Another widely researched concept in positive psychology is 'flow'. Flow is 'the state in which people are so involved in an activity that nothing else seems to matter; the experience is so enjoyable that people do it even at a great cost, for the sheer sake of doing it' (Csikszentmihalyi, 1991/2008, p. 4). In this state of 'optimal experience', one feels 'a sense of exhilaration, a deep sense of enjoyment that is long cherished and becomes a landmark in memory for what life should be like' (ibid, p.3). A lot of research has been carried out coupled with a number of publications to give us an understanding of flow. A good deal of descriptive information is accumulated and some practices are suggested to cultivate the experience of flow. However, much of it is *ad hoc*; I find little cohesive structure in all this research that gives a clear and unambiguous conception of flow. What we have is a conglomerate of subjective feelings with no binding rationale (Rao, 2014a).

In Indian psychology, we have the analogous concept *samādhi*. It is similar to flow, but much more. *Samādhi* has behind it a theory of mind; and systematic methods are suggested for cultivating it. *Samādhi* is not a cognitive state. Rather it is a transcognitive experience which manifests transformative behaviour. Here is a state which defies conventional categorization of experience into cognitive, emotive and conative forms. It is a unitary state of knowing and blending into each other. There is nothing parallel to it in conventional psychological wisdom.

In the Indian tradition, it is possible to control the mind by practising such disciplines as yoga. The control of mind essentially involves ego-transcendence. The ego is pseudo-self that masquerades as the true self and generates illusory identities that create the barriers between 'I' and the 'other'. The pseudo-self is culturally bound and is the conditioning influence on the person. Its needs are not intrinsic to human nature. If Csikszentmihalyi had taken the theoretical model of *samādhi* to construct the concept of flow instead of drawing it from the varied experiences of people from different backgrounds, he would have gone beyond noting the phenomenology of flow to a more coherent theory-laden concept with great potential. It would have helpful in cultivating consciousness with significant outcomes and transformational consequences to people, than is presently the case.

We need to explore positive psychology in India from the perspective of Indian psychology. This would have important implications for enriching positive psychology and extending it beyond the limited Western perspective that currently dominates it. It is in this context I suggest that the present book may be seen as extremely relevant.

The editors of the volume state that the book is prepared from 'the perspective of building efficiency rather than targeting deficiency'. It is intended to provide a wider perspective to the reader by putting forth 'a comprehensive synthesis of Western and Eastern perspectives'. The editors seek to achieve these goals by providing 'comprehensive understanding of general issues of positive psychology' and by examining the applications. It is for the readers to judge the extent to which the editors have succeeded in achieving these goals.

Updesh Kumar, Archana and Vijay Parkash, the editors of this volume, are scientists working at the Defence Institute of Psychological Research. This book is a laudable attempt to focus on the relevance of positive psychology to address current issues in psychology. The wide range of topics deal with Indian inputs in progress of positive psychology, virtues, character strengths and well-being, resilience, happiness, stress and so on. There is a chapter on mindfulness meditation as an aid 'to combat the stress-load of military services'. There is also a chapter attempting to develop an 'analytical model of happiness'.

The book is a welcome addition to literature on positive psychology from India. My perusal of the contents and the rationale behind the book reinforces my belief that Indian psychology in general has great relevance to several contemporary psychological issues and that positive psychology and Indian psychology have much to benefit from each other. If this interesting volume could spur further interest in pursuing positive psychology from the perspective of Indian psychology its publication would have been very rewarding for all those associated with it

References

Csikszentmihalyi, M. (1991/2008). *Flow: The psychology of optimal experience*. New York: Harper Collins.

Diener, E., Lucas, R.E. and Oishi, S. (2005). "Subjective well-being: The science of happiness and life satisfaction. In C.R. Snyder and S.J. Lopez" (Eds.). *Handbook of positive psychology* (pp. 63–73). New York: Oxford University Press.

Kao, H.S.R. and Sinha, D. (1997). *Asian perspectives on psychology*. New Delhi: Sage.

Maslow, A. H. (1954). *Motivation and personality*. New York: Harper & Brothers.

Rao, K. R. (2011). *Cognitive anomalies, consciousness and yoga*. New Delhi: Center for Studies of Civilizations and Matrix Publishers.

Rao, K. R. (2014a). Positive Psychology and Indian Psychology in Need of Mutual Reinforcement. *Psychological Studies* (in press).

Rao, K. R. (2014b). Indian psychology in prospect. *Psychological Studies* (in press).

Seligman M.E.P. and Csikszentmihalyi, M. (2000). Positive psychology: An introduction. *American Psychologist*, 55, 5–14.

K. Ramakrishna Rao, Padma Shri
Chancellor, GITAM University, Vishakhapatnam

About the Editors and Contributors

Editors

Updesh Kumar, Ph.D., is Scientist 'F', and in the chair of the Head, Mental Health Division at Defence Institute of Psychological Research, R&D Organization (DRDO), Ministry of Defence, Delhi. He obtained his doctorate degree in the area of *Suicidal Behaviour* from Panjab University, Chandigarh, India and has 24 years of experience as a Scientist in an R&D organization. He specializes in the area of research in suicidal behavior, low-intensity conflict, personality assessment and terrorism. He has edited five quality volumes on *'Recent Developments of Psychology'*, *'Counseling: A Practical Approach'*, *'Suicidal Behaviour: Assessment of People-At-Risk'*, (2010, published by *Sage Publications*) *'Countering Terrorism: Psychosocial Strategies'* (2012, published by *Sage Publications*) and recently published *'Understanding Suicide Terrorism: Psychosocial Dynamics'* (2014, published by *Sage Publications*) and *'Suicidal Behaviour: Underlying Dynamics'* (2015, published by Routledge). The highlighted four books with Sage Publications and Routledge have been completed in collaboration with internationally acclaimed academicians/researchers from India, USA, UK, Australia, Singapore, Israel, Germany, Hungary, Hong Kong, Norway, Ireland, Netherlands, and Spain. He is the author/ editor of various field manuals for the Armed Forces personnel and general population and almost 50 other articles, research papers and book chapters. He was conferred with the DRDO's Best Popular Science Communication Award-2009 by the Ministry of Defense of India. He has also been the recipient of DRDO Technology Group Award in 2001 and 2009, Professor Manju Thakur Memorial Award-2009 and 2012 for his outstanding and innovative academic publication by Indian Academy of Applied Psychology (IAAP) and Professor N. N. Sen Best Paper Award for the year 2010 by the Indian Association of Clinical Psychologist (IACP). Recently, he has been conferred with Laboratory Scientist of the Year Award-2013 and DRDO Scientist of the year Award-2013 by Government of India.

Archana, Ph.D., is Scientist 'D' at Defence Institute of Psychological Research and is associated with R & D activities in the areas of organizational behaviour, mental health and well-being of soldiers in peace as well as field locations. She has to her credit a number of papers in the national journals of repute. She has authored two manuals on 'Stress Management' and 'Psychological Well-Being of Soldiers at High Altitude'. In addition, she has made valuable contributions in organizing many workshops and seminars for the soldiers in the area of military psychology. She has been the course coordinator in organizing training programs for service officers and has also been actively involved in delivering lectures on stress management and other related areas in active field environments. She was awarded National Technology Day Award - 2014 by DRDO for her outstanding contributions.

Vijay Parkash, Ph.D., is Scientist 'C' at Defence Institute of Psychological Research, Defence R & D Organization, Delhi. After completing his post-graduation as Gold Medallist in the entire social sciences faculty of Kurukshetra University, he was awarded DRDO Research Fellowship and he completed his Doctorate degree in Psychology from Kurukshetra University. Health psychology, personality and psychometrics are the areas of his interest and he has eleven years of research experience. He has also served as a Psychologist at Air Force Selection Board, Dehradun for around two years. He has been involved in many major research projects related to suicidal behaviour and test constructions for Personnel Selection in Armed forces and Paramilitary Forces. He has been one of the editors of two edited volumes—*'Recent Developments in Psychology'* and *'Counseling: A Practical Approach'*; and has more than 15 other academic publications in the form of journal articles and book chapters to his name. Defence Institute of Psychological Research conferred upon him the Scientist of the Year Award in 2011 for his outstanding contributions.

Contributors

Kiran Kumar K Salagame, Ph.D., is a Clinical Psychologist and a Professor of Psychology at the University of Mysore. Integrating Indian psychological concepts to mainstream psychology is his current focus. His research and publications relate to meditation & yoga, states of consciousness, Indian psychology, holistic health, positive psychology, transpersonal psychology and social cognition. He has authored a book on *The Psychology of Meditation: A contextual approach*. Indian Association of Clinical Psychologists conferred 'Psycho Award' in 2006 for his significant contributions to the field of psychology. He was a Fulbright fellow in 2011 and in 1990–91. Currently, he is an Honorary Professor, International Academy for Yoga Teacher Training, Belgrade, Serbia; Member, Scientific Board at International Society for Interdisciplinary Yoga Research, Belgrade, Serbia; He is also a member of the Board of International Transpersonal Association, USA and Academic Council, Swami Vivekananda Yoga University, Bangalore. Kiran Kumar has been serving as the Associate editor of *Psychological Studies* and is on the editorial boards of national and international journals. He is a Fellow of Indian Association of Clinical Psychologists, and member of National Academy of Psychology, India.

Jitendra K Singh, Ph.D., is working as Scientist 'D' in Defence Institute of Psychological Research (DIPR), Defence Research and Development Organization, Ministry of Defence, Delhi since 2004. Before joining DRDO, he has been lecturer of Applied Psychology at VBS Purvanchal University, Jaunpur (U.P.). He completed his M. Phil and Ph. D. from University of Delhi. His main areas of interest are personality, political psychology, perception management, and inter-group relation. He has been contributing in the area of teaching, research and training for more than thirteen years. Singh has participated in more than twenty-five national/international conferences. He has organized an international conference on military psychology and four symposia in different conferences. He has got twenty-one research publications in national and international journals and six book chapters. At present, he is Convener, Division of Military Psychology of National Academy of Psychology (NAOP), India. He received Young Scientist award from NAOP in 1998, Technology Day Oration Award from DRDO in 2010, Technology Group Award by DIPR in 2011 and Laboratory Scientist of the Year Award by DIPR in 2014.

Surendra Kumar Sia, Ph.D., is presently working as an Associate Professor in the Department of Applied Psychology, Pondicherry University, Pudducherry. He has done his M.A., M.Phil., and Ph.D. in psychology from University of Delhi. He has more than 12 years of teaching and research experience and successfully supervised some Ph.D. scholars and M.Phil. students. To his credit, he has 24 research publications in journals of national and international repute. He is a member in the editorial board of Journal of Happiness and Wellbeing, International Journal of Psychological Studies and Indian Journal of Health and Wellbeing. He is also an executive committee member of National Academy of Psychology, India. His areas of interest include organisational behaviour, environmental psychology and positive psychology.

Amrita Yadava, Ph.D., is Professor and Head of the Department of Psychology at Maharishi Dayanand University, Rohtak. Professor Yadava is extensively engaged in teaching, research and application work in Psychology for last more than 30 years. Her primary areas of teaching are experimental and cognitive psychology. She has published more than 85 research articles in various reputed national and International journals. In addition to participation in large number of seminars and conferences, she has edited nine books on important applied fields of Psychology. In current years, she is conducting workshops and skill development programs in schools and colleges on 'Positive Youth Development', 'Gender Sensitization and Female Issues', 'Environment Education' and 'Positive teaching and parenting'. She, being the Director of Women Studies Center at M. D. University, Rohtak has initiated many programs targeting female adolescents' problems, needs and remedies.

Hardeep Lal Joshi, Ph.D., is Assistant Professor, Department of Psychology, Kurukshetra University, Kurukshetra. He was awarded two gold medals for getting first-class first position in the university alongwith highest percentage of marks in the Faculty of Social Sciences at Post Graduation level. He completed his doctoral dissertation from Kurukshetra University. He has more than 15 years teaching as well as research experience. He completed his specialized degree in Clinical Psychology from IHBAS which is affiliated to University of Delhi. His specialization is in the areas of clinical psychology, psychological testing, and mental health. He has guided more than twenty M.Phil. and ten Ph.D. students. He has published twenty-five research papers in journals of national and international repute.

Akbar Husain, Ph.D., is Professor and Coordinator, UGC—SAP (DRS-I) at the Department of Psychology, Faculty of Social Sciences, Aligarh Muslim University, Aligarh. He received his Ph.D. Degree in Psychology from the Aligarh Muslim University, Aligarh (INDIA) in 1981.

He has over 32 years of teaching and research experience. He has successfully supervised 7 Research Reports to the M.Ed. students at University of Malaya, Malaysia, 10 M. Phil dissertations and 19 Ph.D. theses at Aligarh Muslim University, Aligarh. He has authored, co-authored and edited 27 books and contributed 175 research papers, theoretical articles and book chapters. He has been the Editor of the Journal of the Indian Academy of Applied Psychology (2002-2005) and Journal of Educational Research, University of Malaya (2009). Professor Husain is also the reviewer of various National and International Journals of Psychology. He has standardized numerous psychological tests and completed four research projects. His current areas of research are—spiritual psychology, health psychology, counseling psychology, and positive psychology.

Shahin Zehra is a Doctoral Research Scholar and Project Fellow in UGC-SAP (DRS-1), Department of Psychology, Aligarh Muslim University, Aligarh. She has presented many research papers in various national and international conferences and has also published some papers in refreed national journals.

Papri Nath, Ph.D., is Senior Research Fellow in Department of Humanities and Social Sciences, Indian Institute of Technology, Kharagpur. She received the 'Mother Theodosia A.C. Memorial Scholarship' for securing first rank in Psychology (Bachelors) at the university level. She is the recipient of the 'Junior Research Fellowship' granted by University Grants Commission. Before joining IIT Kharagpur, she has taught courses in psychology, stress management and industrial psychology as a lecturer at Patna Women's College. Her areas of research interest are positive emotions, entrepreneurship (micro-approach), emotional intelligence, emotional labour, aging and retirement, individual and organizational well-being, management of workforce diversity. She has published a significant number of research articles in reputed national and international journals.

Rabindra Kumar Pradhan, Ph.D., is Associate Professor of Psychology at Department of Humanities and Social Sciences, Indian Institute of Technology, Kharagpur. He has earlier been a Scientist in Defence Institute of Psychological Research, Defence Research and Development Organization, Ministry of Defence, India. His areas of specialization include social and organizational psychology, industrial psychology, managerial psychology, organizational behaviour, human resource development and management. He has more than 15 years experience in the field of training, teaching and research in the above fields. Pradhan is a member of various professional bodies, including International Association of Applied Psychology, International Association of Cross-cultural psychology, Indian Society for Training and Development, National Academy of Psychology, India, Indian Academy of Applied Psychology and Indian Science Congress Association. He has to his credit four books and more than 30 other publications in referred journals, newspapers, magazines, and books.

Rajbir Singh, Ph.D., is Senior Professor at Department of Psychology and currently the Dean, Students' Welfare at Maharishi Dayanand University, Rohtak. He has earlier served as Dean, Faculty of Social Sciences and Head, Department of Psychology at Maharishi Dayanand University, Rohtak; and, Chairperson, Department of Psychology, Kurukshetra University, Kurukshetra. He specializes in the area of positive psychology, psychobiology and neuropsychology. He has been the founder-president of Haryana Psychological Association and is affiliated to many national and international academic bodies as a life member. He is the Coordinator of UGC-SAP-II project on Health Psychology, subject matter expert for UGC examinations, invited expert for Union Public Service Commission and member of academic bodies of many universities. He is the founder editor of the Journal of Indian Health Psychology, has authored 6 books and has published more than 50 research papers in national and international journals.

Lokesh Gupta, Ph.D., has done his M.Phil. and Doctoral Research at Department of Psychology, Maharishi Dayanand University, Rohtak. He has been a Research Investigator in Defence Institute of Psychological Research, Delhi and Project Assistant at Department of Humanities and Social Sciences, Indian Institute of Technology, Delhi. His areas of interest are psychometrics and positive psychology. He has presented papers in conferences and published research papers in refereed national journals.

Suvashisa Rana, Ph.D., is an Assistant Professor at the Centre for Health Psychology, University of Hyderabad. He has been the recipient of the University Gold Medal from Utkal University and the coveted Doctoral Fellowship of Government of India by the Bureau of Police Research & Development, Ministry of Home Affairs, Government of India to pursue his PhD research on Police–Public Partnership for Peace. He has served as Psychologist under Government of India after being selected by the Union Public Service Commission for a substantial period of time. His current area of research is positive psychology, health psychology and psychometrics.

Meena Hariharan, Ph.D., is Professor in Health Psychology in University of Hyderabad. She is the Founder-Director of the Centre for Health Psychology, University of Hyderabad. She has been working as faculty in this University since 1992. Prior to this she worked in the National Institute of Nutrition, Hyderabad. She has three books and about 60 articles to her credit. Her research areas are coping with stress, icu trauma, psychosocial factors in non-communicable diseases. She has completed 30 research projects funded by national and international agencies.

N K Chadha, Ph.D., is Professor and Head of the Department of Psychology, University of Delhi. Professor Chadha is the former Chairman, Board of Research Studies in Humanities and Social Sciences, Delhi University and former Head, Department of Adult, Continuing Education and Extension, University of Delhi. Professor Chadha has been teaching in the Department of Psychology since 1982. Professor Chadha has authored several books including *Applied Psychometry, Aging and the Aged-A challenge to Indian Gerontology, Recruitment and Selection, Organizational behavior, Human resource Management, Be a winner,* and *Social Psychology*. He has also coauthored a book with John Van Willigen from University of Kentucky, USA entitled 'Social Aging in Delhi Neighborhood'. He has been conducting many ongoing research projects with foreign universities and has carried out a number of major interdisciplinary and collaborative projects. Professor Chadha is the President of Indian Association of Positive Psychology and India Career Development Association and has published more than 115 articles in various national and international journals.

Vandana Gambhir, B.H.M.S., a research scholar of the Department of Psychology, University of Delhi, is an alumnus of Nehru Homoeopathic Medical College and Hospital, one of the premier and reputed homoeopathic colleges of India. Started her professional career as a physician, she was also a faculty member and content developer for leading training institutions for the Civil Service and Indian Forest Service Examinations. Her field of involvement in psychology research-work span across psychometric testing, organizational behavior, personality psychology, competency mapping, positive psychology, consumer behavior and geriatrics mental health.

Nilanjana Sanyal, Ph.D., is a Professor in the Department of Psychology, University of Calcutta, Kolkata. She has been practicing psychodynamic psychotherapy and psychoanalysis for the last 31 years. She imparted training in counselling in various governmental institutions and N.G.O.s, both nationally and internationally, and is consistently involved in mass media programmes for awareness-generation regarding mental health issues. She is a Visiting Professor at Department of Adult and Continuing Education, Jadavpur University, Department of Psychology, Dhaka University, and Department of Applied Psychology, Pondicherry University, Mizoram University and Central Institute of Psychiatry, Ranchi. She has to her credit a number of national and international publications and is also a columnist in different national dailies and magazines. Her research interests include personality and clinical psychology, interpersonal relationships, and psychodynamic psychotherapy.

Sibnath Deb, Ph.D., is Professor and Head of the Department of Applied Psychology, Pondicherry University, Pudducherry, India and Adjunct Professor, School of Public Health and Social Work, Faculty of Health, Queensland University of Technology, Brisbane, Australia. He has 25 years of teaching and research experience. In 1994 Professor Deb did an intensive course on 'HIV/AIDS and Qualitative Research' from the University of Western Australia. Prof. Deb has guided ten Ph.D. scholars, conducted more than 50 large-scale qualitative and quantitative studies (funded by national and international funding agencies), and published more than 100 research papers in addition to writing eight books and editing four books. He is on the editorial board of the '*Journal of Interpersonal Violence*', published by the University of Washington, USA. He has received three international and three national awards for his contribution in the field of child safety and applied psychology. His current areas of research interest include child safety/protection, students' mental health, adolescent reproductive health, HIV/AIDS and applied social psychology.

Aneesh Kumar, M.Sc. in Applied Psychology with specialisation in Clinical Psychology from the Pondicherry University, has been carrying out his doctoral research in the Department of Applied Psychology, Pondicherry University. Currently, he is working as Project Officer in a research project on Burden of Child Abuse and Neglect, funded by the Indian Council for Social Sciences Research (ICSSR). He has also done his P.G. Diploma in Industrial Psychology from Pondicherry University and has published five research papers and attended five conferences.

Anjali Gireesan, M.Sc. in Applied Psychology with specialization in Clinical Psychology from the Pondicherry University, has been carrying out her doctoral research in the Department of Applied Psychology, Pondicherry University. Currently, she is working as Project Fellow on a research project on Academic Stress of School Students, funded by the University Grants Commission, New Delhi. She has published seven research papers and attended five conferences.

Nov Rattan Sharma, Ph. D., is a Professor of Psychology at M. D. University, Rohtak. He has a teaching and research experience of more than 30 years. His area of research is personality and positive health. He has successfully supervised 17 Doctoral theses and published more than hundred articles in various books and journals of repute. He has also edited ten books on different topics of applied psychology. He is the editor of a *The Journal of Indian Health Psychology.* He is the life member of many national and international professional and academic associations. Professor Sharma is actively engaged in DRS-II (SAP, UGC) program of the department where awareness, assessment and promotion of health is the key issue of the extension services. He has received two prestigious awards from the Indian Academy of Applied Psychology (Prof Mrs. Manju Thakur Memorial Award 2011 and Prof Anima Sen Award 2013).

PART 1
Theoretical Perspectives

CHAPTER 1

Indian Perspectives and Positive Psychology

Kiran Kumar K. Salagame
University of Mysore, Karnataka

Abstract: Indian traditions, 'Vedic' and 'non-Vedic', have extensively dealt with issues of human happiness and well-being which are the core themes of contemporary positive psychology. Since the Indian traditions understand human nature and also the nature of the universe in a more integrated fashion as experienced from a state of consciousness that transcends the space–time dichotomy, their perspective on the nature of reality emerges from a higher state of awareness which is beyond voluntary consciousness and their understanding of the subject matter has greater depth and breadth as compared to that of contemporary perspectives. Though there are differences in these traditions, they converge on the point that there is more to reality than what meets the eye. Among the many aspects discussed about the nature of reality, consciousness, human nature, life, death and so on, the issues of happiness and well-being takes the centre stage, because it is observed that ultimately everyone is concerned with them. While in the Vedic tradition, happiness is approached directly as a state that is intrinsic to human nature itself as in Sat-Chit-Ānanda, Buddhist tradition discusses about it indirectly by focusing on suffering. These two primary approaches have led to two major perspectives on the nature of well-being and associated practices to reach the goal. In view of this, we are justified in speaking of Indian perspectives on positive psychology.

Indian Perspectives and Positive Psychology

Positive psychology rings a bell in the Indian psyche of familiarity. Happiness and well-being brings to mind so many words, like *ārogya, sukha, samtosha, tripti, tushti, harsha, ullāsa, swāsthya, ānanda* and so on (Salagame, 2006c). Similarly, altruism, courage, forgiveness, gratitude, spirituality, transcendence, wisdom and so on brings up to mind many familiar terms *niswārtha, dhairya, kshama shīla, krtajñata, ādhyātma, prajña* and others.

When I read some of the initial research studies related to the limitations of wealth, sensory pleasure and exclusive pursuit of individuality, independence and autonomy to the neglect of relationships; about the need for development of family bond; about how satisfying it was for Seligman and his wife to allow their children in their infancy and early childhood to sleep with them in their bedroom sacrificing their sense of freedom; how breast feeding is psychologically very important; and many such findings (Seligman and Csikszentmihalyi, 2000; Seligman, 2002). My immediate gut reaction was that all these are very familiar to me as grandmothers' wisdom in my country! I felt so, because in our culture many of them have been spoken about in a matter of fact way in folklore, in literature, in poetry and in religious and philosophical texts derived from conventional wisdom. However, beyond this initial gut feeling, my academic and conceptual response to these developments was to view the emergence of positive psychology as an emphasis to move away from exclusive preoccupation with *artha* and *kāma* as goals of life; and as a shift in focus, from the preoccupation of modern psychology with *rajo guna* and *tamo guna* towards studying more of *satva guna* (Salagame, 2002a, 2002b, 2003).

> In the last decade, more and more research findings have vindicated this view of mine. It appears that much of Indian psychology is devoted to the development of *satva guna* and, hence, it is not wrong to consider Indian psychology and positive psychology as 'birds of the same feather that flock together' (Salagame, 2006b). I have discussed elsewhere, how we can understand the notion of happiness and well-being from Indian perspectives, in terms of concepts like *triguna, panchokosha, purushārhta, preyas and shreyas, sukha and dukkha, swāsthya,* etc., (Salagame, 2014).

Not only did the ancient Indian thinkers analyse the nature of happiness, well-being, etc., threadbare, but also did they find ways of achieving a sense of well-being that is sustainable and termed it as *ānanda* and distinguished it from *sukha* which is conditioned by the spatio-temporal limitations of life. That is why, in our country, spiritually realized persons have their names suffixed or prefixed with *ānanda*, as in Swāmi Vivekānanda, Swāmi Shivānanda, Parmahamsa Yogānanda, Mā Ānandamayi, Māta Amrtiānandamayi, and so on.[1]

Jane Henry (2013: 421) notes that Eastern mysticism 'documents many different higher states of consciousness which often leave the recipient in a state of joyful, perceptive, equanimity'. She observes, 'western notions of happiness tend to be associated with high arousal, whereas the contentment and equanimity found in spiritual practice represent states of well-being associated with lower arousal' (ibid.). In her opinion psychologists 'could usefully spend more time examining different states of optimal experience and well-being; in particular, those associated with low arousal such as contentment' (ibid.). We seem to be shifting gears from

[1]The renunicates of the Ramakrishna Mission and many other orders of monks in India have their name suffixed with *ānanda* mandatorily.

low arousal to high arousal in our culture, as a whole; moving towards *artha* and *kāma*; and becoming more *rājasic* and *tāmasic* losing much of *sātvic* nature.

Positive Psychology: A Very Short History, But a Very Long Past

As the popular English saying goes, 'positive psychology has not just short history, but a very short one and yet a very long past. Positive psychology just entered its teenage period last year, since its formal launch with a special issue of *American Psychologist* in January 2000. However, its growth has been very fast and has already made a name for itself like a child prodigy! But its identity formation continues. However, it has very long past goes as far back as human history itself, when man could conceptualize good and bad, pleasure and pain, happiness and sorrow, etc. As Seligman and Csikszentmihalyi (2000) notes, 'modern psychology focused too much and too long on the negative side of the polarity on anxiety, anger, depression, etc., and it is time to shift our focus on happiness, joy, contentment, and so on'. Henry (2013: 411) states, 'Psychology has only recently turned its gaze to investigate happiness and well-being while spiritual traditions have been scrutinizing this for millennia'.

Thus, the agenda of focusing on *positive subjective experiences*, *positive individual traits*, and *positive institutions* (Seligman and Csikszentmihalyi, 2000) caught the imagination of many psychologists with a proliferation of thousands of research articles and hundreds of books. It became a rallying point for many researchers who were already working on these issues independently that appeared as seemingly disconnected areas till they were brought together under one title (Ryan and Deci, 2001). Since then, three key themes, viz., *optimal psychological functioning* (Snyder and Lopez, 2002); *happiness and well-being* (Seligman, 2002); and *virtues and character strength*s (Peterson and Seligman, 2004) have constituted the core subject matters of positive psychology. Yet defining them precisely has remained a daunting task. Nevertheless, there is a rapid progress in understanding the nature, varieties, conditions and underlying psychological processes involved in them (Sheldon, Kashdan and Steger, 2011).

What is striking about the outcome of a decade of research in positive psychology is the growing realization that whatever was once thought by the West as the contributing factors to attain satisfaction in life and experience a sense of well-being are really not so. That list primarily includes earning more money, pursuing sensory pleasures, and attaining individuality to the neglect of relationships. It is observed that these factors lose their capacity to bring satisfaction and sense of well-being beyond a limit.

The emergence of positive psychology in USA is admittedly a cultural phenomenon and has emerged in response to the value crisis experienced in that society (Seligman, 2002). There are two aspects to this development. One is the recognition of the fact that any amount of material richness will not proportionately increase the felt sense of happiness or well-being among people. This is very well brought out in a book titled, *The Poverty of Affluence: A Psychological Portrait of the American Way of Life* by Paul Wachtel (1989), which provides a diagnostic insight from a trained clinician to the American situation. Many economically poor countries are found to be at par with the rich countries on the index of life satisfaction even though their purchasing power parity is quite uneven. For example, one study (cited in Seligman, 2002) shows that USA with Purchasing Power Parity (PPP) of 100 on a scale of 1–100, has a life satisfaction score of 7.75 on a scale of 1–10. The same study shows that India has a life satisfaction score of 6.70 even though PPP is only 5. Just a difference of a single point on the life satisfaction index for a difference of 95 points on PPP! There are many other studies that report similar trends (Diener, 2000; Diener and Diener, 2002; Diener and Seligman, 2004). What does it tell? We need to ponder.

The second aspect is the moral and ethical value crisis experienced in that society resulting in focusing on the study of virtues and character strengths (Peterson and Seligman, 2004), which were once considered as value-loaded terms and, hence, not scientific enough to deal with in the mainstream psychology. It is in this background that researchers have focused on many other topics which were not considered as of much psychological significance. That list includes altruism, forgiveness, gratitude, love, courage, creativity, meaning, wisdom, spirituality, transcendence; all those traits humanistic psychologists associated with self-actualization.

More and more researchers are recognizing that human life is more fulfilling with the development of virtues and character strengths, healthy interpersonal relationships and living in supportive community. Barbara Frederickson's 'Broaden and Build Theory' shows how love and meditation can enhance positive effects. Bourgeoning literature on the positive effects and benefits of mindfulness meditation shows how erroneous it was for psychoanalysts to view such practices as regressive in nature. Research on the positive outcome of gratitude exercises is another important focus. Last, but not the least, some researchers observed that happiness is intrinsic to human nature and seems to depend less and less on external circumstances.

While the literature related to happiness primarily focuses on *hedonia*—the pursuit of pleasure and positive effects, the age-old Greek concept *eudaemonia* is about focussing on the studies related to virtues and strengths, optimal psychological functioning, positive relations, and positive institutional building. These two have been associated with two conceptions of well-being, viz., *Subjective Well-being* (SWB) of Diener (1984) and *Psychological Well-being* (PWB) of Ryff (1989). Study of the relationship of happiness to stages and conditions of life across the lifespan and to variations in values across cultures are other important focus of research in positive psychology (Baumgardner and Crothers, 2006). The question that is actively debated is, whether *hedonia* and *eudaemonia* represent two different kinds of happiness and what are the problems in distinguishing them (Kashdan, Biswas–Diener, and King, 2008; Kristjansson, 2010). Indian thinkers have addressed many of these issues (Salagame, 2013a; Sinha, 1985). However, the way Indian perspectives address the themes of positive psychology, differs significantly.

A Note on the Indian Perspectives

A perspective is what provides a viewpoint on anything, be it something as natural and concrete as viewing one's physical environment from a mountain top as against watching the same from the foot of the mountain; or something as highly abstract and conceptual as the fundamental nature of reality or truth, of consciousness, life and death, etc. In India, we do have many viewpoints, thus justifying the usage 'Indian perspectives', which seem to be the bases for the enormous diversity that exist in our country in all aspects of culture including caste, creed, language, food habits, customs, traditions, rituals, festivals, and so on. This enormous cultural diversity at both abstract and concrete levels, including the natural diversity in flora and fauna, has made foreigners to exclaim, 'Incredible India!' which has become a slogan for tourism promotion in our country.

In Sanskrit, the term 'perspective' is known as *darśana*. As Hiriyanna (1993: 182) a well-known Indian philosopher notes *darśana* is the 'general name for the results arrived at by means of the several *pramānās*. *Pramānās* are the means of obtaining valid knowledge such as perception (*pratyaksha*), inference (*anumāna*), experimentation (*prayoga*), authority (*shabda*), and others, which are discussed by the branch of philosophy known as epistemology.' Hence, *darśana* means, 'philosophic opinion' and it signifies a specific school of thought as distinguished from others. Thus, Indian philosophical systems are referred to as *darśana*. It is a convention in India to regard Vedic paradigm as orthodox. Therefore, *darśana* is divided into two categories *āstika* (orthodox) schools and *nāstika* (heterodox) schools.

Hiriyanna (ibid.) notes that the distinction made between orthodox and heterodox is arbitrary and, hence, it is not absolute, but relative. Further, the terms *āstika* and *nāstika* are used as synonyms for theism and atheism by many scholars and researchers. This is a wrong usage, because the two Indian terms have altogether different connotations. As Śastri (1997) notes, '*āstika* denotes the systems which recognize the Vedas and their branches as the supreme authority'. It does not denote 'theism' in the Western sense. *Sāmkhya*, for instance, is an atheistic philosophy, yet it is regarded as *Brāhmanic* system (a *āstika* system) since it has accepted the authority of the Vedas. But, Buddhism and Jainism are considered to be non-*Brāhmanic* (*nāstika* systems), because they do not recognize the authority of the Vedas. According to another interpretation, *āstika* is one who believes in the existence of the future world, etc. As per this interpretation, the Buddhists and the Jains cannot be called *nāstikas*, because they believe in reincarnation. So, what distinguishes these two types of systems is whether they owe allegiance to Vedas (*vaidika*) or not (*avaidika*). Otherwise, they tend to hold similar views on certain fundamental assumptions of ontological nature (Salagame, 2008a).

There are six *āstika* schools and six *nāstika* schools. *Āstika* schools are the *Nyāya* of Gautama, Vaiśeşika of Kanāda, Sāmkhya of Kapila, and Yoga of Patañjali, Pūrva-Mimāmsa of Jaimini and the Uttara-Mimāmsa or Vedānta of Bādarāyana. Traditionally, scholars have grouped Nyāya-Vaiśeşika, Sāmkhya-Yoga and Pūrva-Mimāmsa, Uttara-Mimāmsa or Vedānta together, forming three pairs, because they have found that the schools forming each pair agree either in their general metaphysical outlook or in their historical basis or in both. The ontological assumptions underlying the different *darśana* are provided in a tabular form (*see* Table 1.1).

According to Hiriyanna (1993), another meaning of the term *darśana* is in the sense of its equivalent term *drişti*, which literally means 'sight' and may be taken to indicate that what the Indians aspired after in philosophy was not a mediate knowledge of the ultimate truth, but a 'direct vision of it'. Such a direct vision of truth is termed as *yogi pratyaksha* and *swayam prakāshatva*. The word *darśana*, thus, conveys the difference between modern Western philosophy, which mainly relies on intellectual pursuit and Indian philosophy that aims

Table 1.1 Ontological Assumptions Underlying the Different *Darśana*

Ontology	Explanation	Systems
Materialists	Those who hold that reality is only material as perceived by sensory organs and there is no metaphysical reality.	*Cārvāka* or *Lokāyata*
Realists	Those who believe in the existence of objects independent of mind and the *Atman*.	*Mimāmsaka, Nayāyika, Vaiśeşika, Jaina* and *Bouddha*
Pluralists	Those who believe that there is a plurality of objects and *jivas* and all of them are real.	*Mimāmsaka, Nayāyika, Vaiśeşika, Jaina*, and *Bouddha*
Particularists	Those who hold that all particulars are independent of one another and they are infinite in number and cannot be reduced to anything in common.	*Vaiśeşika*
Idealists	Those who believe outward reality has its base in an inner reality.	*Sāmkhya, Yogi*, and *Vedānta*
Unitarists	Material world which appears plural is derived from a unitary principle.	*Sāmkhya*
Monists	Those who hold that there is no plurality and reality is one.	*Advaita, Vishishtādvaita*

Adapted with permission from Raju, 2009.

at transforming the insight and conviction arrived through such intellectual pursuit into direct experience (Salagame, 2008a). To borrow a term from Abraham Maslow (1971), what our ancient thinkers aimed at was 'Being-cognition'. Systems of Indian thought have developed either accepting the Vedic paradigm (*āstika*) or independent of it (*nāstika*). Vedas are said to be products of revelation and intuition and the Vedic rishis were cognizers of the Vedic hymns (*mantra drastāra*) (Salagame, 2008a). Even non-Vedic traditions were founded by persons who had a direct vision of truth, such as Tirthankaras of the Jaina tradition and Gautama Buddha.

Whether Vedic or non-Vedic, all Indian perspectives emphasized on knowing or realizing the truth, directly or unmediated. In the Vedānta system, the direct, unmediated cognition is referred to as *aparokshānubhuti* and is contrasted with sensory mediated or indirect knowledge, *paroksha.* Hence, Indian and Western perspectives are distinguished on epistemological grounds. While the former admits intuition and revelation as valid sources of obtaining knowledge the latter has come to limit itself to sensory perception and inference supported by experimentation in modern times. There is no place for intuition and revelation in Western tradition and contemporary cognitive psychology does not discuss these human capacities.

Darśana: Higher Stages of Cognitive Development and Paradigms

Indian philosophers like Sri Aurobindo, Hiriyanna, Jwala Prasad and others have observed that development of Indian thought traditions underwent different cognitive phases beginning with a predominance of intuition during the Vedic period and then gradually moving towards predominance of intellect during the Upanishadic period culminating in the development of *darśanas* based on logic and reasoning (Salagame, 2008b). Their observations suggest that there lies a necessary link between our cognitive systems and the worldview that we hold.

Jean Piaget discussed the relationship between epistemology and psychology under the heading 'Genetic Epistemology'. In recent years, there have been many studies that examine the relation between psychology of individual differences and knowledge acquisition process under the heading 'Personal Epistemology'. These studies show how individual cognitive development plays a vital role in understanding different subjects in one's educational career.

In a similar way, it is possible to relate modes of cognition and stages of cognitive development to different knowledge systems that exist across cultures. It has been suggested that two primary modes of consciousness, viz., intuition and intellect have been differentially emphasized in the Eastern and Western cultures (Ornstein, 1972). Wilber, Engler and Brown (1986) distinguish pre-personal, personal and transpersonal stages of development and relate them to different psychological systems. Swāmi Ajaya (1983) discusses Piagetian and post-Piagetian stages of cognitive development and their relation to four paradigms, viz., *reductionist, humanistic, dualistic* and *monistic*, underlying different psychologies and psychotherapies of the East and West. According to him, the reductionist paradigm is rooted in Piaget's formal operational thought. Dualistic and monistic paradigms emerge from two further stages beyond the formal operational stage. Two post-Piagetian researchers in developmental psychology, Klaus Reigel and Herb Koplowitz, have described these stages. Klaus Reigel describes 'dualistic though' and Koplowitz goes beyond this and describes 'unitary thought'. These two kinds of thought underlie the dualistic and monistic paradigm described by Ajaya. Sāmkhya and Advaita Vedānta are cited as representative of these two paradigms from the Indian tradition. Therefore, all these suggest that Indian indigenous perspectives and concepts have a different foundation than modern psychology (*see* Salagame, 2011, for a detailed discussion).

Indian Perspectives

What implications such fundamental differences between Indian and Western perspectives have for our understanding of positive psychology is our immediate concern? Among the Indian perspectives, it is the position of Cārvāka which is parallel to the reductionist paradigm. They were the modern equivalents of positivists. Since they adhered to the epistemological position that there is no reality beyond what is perceivable through sensory modality, they denied all metaphysical understanding prevalent in the Indian culture. For them, bodily death is the end of life. They said nothing remains except ashes after cremation. No soul, no afterlife. So, they advocated that one should enjoy life thoroughly maximizing pleasure and minimizing pain. But, their position never gained prominence in the Indian culture.

On Pleasure and Pain

The Vaiśeşika hold a position similar to that of humanistic paradigm, but not the same, because all humanistic psychologists do not accept the notion of spirituality and transcendence except Abraham Maslow. As mentioned in Table 1.1, Vaiśeşika are particularists, i.e., they hold that all particulars are independent of one another and they are infinite in number and cannot be reduced to anything in common. The term, 'Vaiśeşika' is derived from *visesha,* that which differentiates. From this, the term *vaishistya* is also derived which means uniqueness. Kanāda (300, BCE), the founder of *Vaiśeşika darśana* holds that pleasure and pain are two irreducible feelings and are different from each other, since they arise from cognitions of either a desirable or undesirable object and are hostile to each other. Many thinkers (Prasastapada, 400 CE, Vachaspati Misra, 900 CE, Sridhara, 1000 CE, Udayana, 1050 CE, Kesavamisra, 1300 CE and Annambhatta, 1700 CE) hold that pleasure is an agreeable feeling. Its nature is one of gratification or satisfaction. Pleasure arises under certain conditions: (1) Proximity of desirable objects; (2) Interaction of the objects with the appropriate sense organs; (3) Conjunction of the self with the mind; (4) Perception of the desirable object; (5) Merit; and (6) Health of the organism. This gives rise to organic (i.e., bodily) expressions like brightness of the eye, beaming of the face, etc., (Sinha, 1985).

It should be noted here that Indian thinkers bring in two conditions—merit (*dharma*) and health (*swasthata*)—as additional factors in the experience of pleasure. Merit or virtue is the peculiar trait of a character acquired by the self due to its past moral deeds and it is the subjective, moral or predisposing condition of pleasure, whereas external objects are the exciting conditions (Sinha, 1985).

Whether one accepts the notion of past moral deeds or not, the fact remains that we cannot enjoy or derive pleasure purely because external conditions are present unless we also have the necessary mental set. This explains why many people are unhappy despite all the riches they have. It is noteworthy that Indian thinkers also considered health as an important condition for experiencing pleasure because we all know that a condition of ill health can hinder enjoyment.

Another important feature recognized is that pleasure or pain is independent of the object, because the same object produces pleasure in one person and pain in another or in the same person at different times (Sinha, 1985). Intra-individual and inter-individual variations in the potentiality of objects, events, etc., in causing pleasure or pain have made Indian thinkers define pleasure and pain with respect to a person's state of mind, thereby emphasizing that it is intrinsic rather than extrinsic.

(A) Kinds of Pleasure and Pain

In distinguishing the kinds of pleasure, Indian thinkers differentiated between pleasure and happiness. Prasastapāda classifies them into four, viz., sensuous, retrospective, prospective and happiness. Sensuous pleasure arises from the perception of desirable objects in interaction with the sensory organs. Retrospective pleasure stems from recollection of previous enjoyment and prospective pleasure comes from anticipation of occurrence in future. While these three kinds spring from the gratification of desires, happiness arises from the conquest of desires. For example, the term *jina* means a hero who has achieved mastery over oneself by conquering desires and negative emotions. From the term *jina* the noun 'Jaina' has been derived and the teachings associated with are known as Jainism. Therefore, it is considered superior and it arises from wisdom (*vidyā*), tranquillity (*shamā*), contentment (*samtosha*) and a peculiar merit (*dharmavisesa*). Wisdom refers to true knowledge of the self. Tranquillity refers to perfect self-control. Contentment is the absence of desire for more than bare necessities of the preservation of life. Merit refers to the highest excellence of virtue which makes the self completely independent of objects of enjoyment. Thus, happiness is rational and abiding, whereas pleasure is physical and temporary (Sinha, 1985).

With regard to pain also, Indian thinkers (Kanāda, Prasastapāda, Gautama, Vātsyāyana, Udayana, Varadarāja, Annambhatta, Patañjali, Prabhakara, Sridhara, Sankara, Ramanuja, etc.) have made similar analysis. Pain is regarded as a disagreeable feeling produced by an undesired object and involves a feeling of being thwarted. Pain occurs due to the proximity of undesirable objects interacting with the sensory organs in conjunction of the self with *manas* associated with demerit. Pain manifests in the form of gloomy appearance (*vicchāyata*) or depression (*dainya*) (Sinha, 1985).

Different schools of thought have classified pain differently. The classification provided by *Nyaya-Vaisesikha* and *Sāmkhya*-Yoga schools have been by far the most frequently used classification, viz., *ādhyātmika*, *ādhibhoutika* and *ādhidaivika*. *Ādhyātmika* refers to self and it is divided into two categories, viz., physical and mental. Physical refers to the bodily disturbances causing uncomfortableness and pain. Mental pain is associated with desires, anger, greed, fear, delusion, dejection, envy, etc. Mental agitations, emotions and passion are not necessarily dependent on external conditions. *Ādhibhoutika* refers to pain and suffering caused by external circumstances of physical nature that include natural calamities, as well as suffering caused by animals and other human beings. *Ādhidaivika* refers to pain and suffering caused by imperceptible agents, ghosts, demons, and other supernatural entities (ibid.). So far so good, and some of these explanations make sense from a reductionist paradigm as well. However, Indian thinkers have gone further in the understanding of pleasure and pain. They explained neutral feeling or feeling of indifference, *upeksha*. They also described the relation between them and even transcendence of pleasure and pain.

(B) Relation of Pleasure and Pain

Indian thinkers dwelt on the question of the relation between pleasure and pain. Though both pleasure and pain are recognized as two independent feelings in their own right, they are considered to be inseparably related to each other. This relation is explained in terms of a sequential occurrence. Sinha summarizes it as, 'A person erroneously regards pleasure as the supreme end of life, pursues it with undivided attention and inevitably comes to grief. Pleasure brings pain in its trail as an inseparable correlate....They are always related to each other. Pleasure is always experienced as related to antecedent or consequent pain. Pain is always experienced as related to antecedent or consequent pleasure....Pleasure alone cannot be pursued and pain alone cannot be shunned' (1985: 71). This type of thinking is known as *cyclical causality* which is a characteristic of dualistic paradigm (Ajaya, 1983).

(C) Upeksha: Feeling of Indifference or Neutral Feeling

Indian thinkers discussed about the possibility of a feeling tone which is neither positive nor negative. But the way it is interpreted, differ from one school of thought to another. The Buddhist recognizes neutral feeling as a distinct kind of feeling. According to Buddhaghosha, a Buddhist scholar, neutral feeling (*upekkhā*) is neither pleasure nor pain. It is not the mere absence of pleasure or pain. It is a positive feeling or experience of what is contrary to both the desirable and the undesirable. A Vedāntic scholar, Vidyāranya, a follower of Samkara, also distinguishes feeling of indifference from pleasure and pain. While Buddhist *upekkhā* indicates absence of any feeling, the neutral feeling referred to by Vidyāranya appear to represent a state with a positive connotation. 'Neutral feeling lies midway, between the two feelings. It is the phenomenal appearance of the intrinsic bliss of the self' (Sinha, 1985: 78). To appreciate this, we need to move further away from individualized consciousness trapped in the *samsāra*.

(D) Transcending Pleasure and Pain

The idea of transcendence in Indian traditions is integrally related to the notion of non-identification and detachment. It was noted that pleasure and pain occurs due to desirable and undesirable conditions since they evoke agreeable and disagreeable feelings. Human beings have a tendency to identify with the former and disown the latter. Therefore, according to Indian thinkers, the trick lies in not identifying with or developing attachment to anything, either positive or negative, which means giving up both attraction and aversion. In the Bhagavad Gita (Chapter 2), Krishna calls such a person *sthitaprajña* and describes the characteristics in terms of how such a person acts or behaves as requested by Arjuna. In those descriptions, we come across the role of desire in causing negative emotions and losing one's equanimity of mind. These verses also describe, how a *sthitaprajña* reacts to pleasure and pain (*sukha* and *dukkha*) and how he relates to everyone and everything in life. They also propose an associative theory of frustration and aggression leading to mental imbalance. They describe, how a *sthitaprajña* regulates one's senses and emotions and is always guided by intellect. These descriptions of a *sthitaprajña* provide a different kind of subjective well-being of Indian variety as contrasted with Diener's variety of subjective well-being, which operates on satisfaction–dissatisfaction dimension. A *sthitaprajña* on the other hand is one who has transcended the opposites in life (verses 55–72).

It is only in the context of dualistic and monistic paradigms that we can make sense of neutral feeling or relation of pleasure and pain and their transcendence.

Dealing with Polarities: Western and Indian Approaches

Ajaya (1983: 41–42) notes, 'everything that exists is part of a polarity and, thus, has its counterpart. The world of ordinary experience is made up of innumerable polar qualities, such as inside/outside, hot/cold, near/far, light/dark, high/low, stop/go, and so on.' According to him, 'the reductionist, dualistic, and monistic paradigm each understand and deal with polarities in a different way.' While reductionist paradigm approaches them as *antithetical*, dualistic paradigms approach those as *complementary* and monistic paradigm emphasize on *transcending* them.

Commenting on the way, Western culture has dealt with polarities, Ajaya observes that throughout its history Western civilization has regarded two sides of a polarity as *antithetical* or opposing one another. Therefore, 'the conception of opposition is ingrained in the Western mind and reveals itself in all aspects of one's life,' and one typically tends to identify with one side of an antithesis and to disown the other. Thus, a person seek to be 'successful rather than a failure'; 'rich rather than poor'; 'happy rather than unhappy'; 'healthy rather than sick'; 'courageous rather than fearful'; and so on. 'The conception that polarities are made up of opposing forces is also expressed in shared belief systems', and this is 'particularly evident in Western religion, in which good and evil are set against one another and regarded as clashing and irreconcilable forces' (1983: 41). He also states that viewing polarities as antithetical is a characteristic more prominent at the stage of cognitive development that comes to ascendency in adolescence, which Piaget described as the stage of formal operations (ibid.: 43).

Note that Seligman himself first came up with the concept 'learned helplessness' and at a later stage of his career he also came up with the concept 'learned optimism', which are opposites. Even the very emergence of positive psychology itself can be seen as a movement from the negative to positive pole of human nature. The distinction of positive–negative, hedonia and eudaemonia, SWB and PWB and their discussion in literature as though they are antithetical polarities is another outcome of a reductionist approach.

The other way of dealing with polarities involves experiencing the opposite qualities as complementary, which may transpose themselves, i.e., transform themselves into one another. This is being expressed in myths and fairy tales wherein small creatures are depicted as transformed into humans; a servant becomes a king; Cinderella is transformed from a lowly servant to a princess; a good fairy turns out to be a witch in disguise and so on. Complementary views are found in more primitive thinking, in myths and dreams, in mystical thought, in a few schools of modern psychology, and in Eastern philosophy, and in both 'more elementary and more evolved stages of cognitive development than in the stage of formal operations, which characterizes Western reductive thought, (Ajaya, 1983: 42). Formal operations thinking, in fact, seem to be an intermediate stage in which the view that polarities are antithetical reigns supreme. But 'at both, less and more evolved stages of cognitive development, one experiences the two sides of a polarity as complementary, interchangeable, and mutually sustaining' (1983: 42–43).

Ajaya's observations about the presence of complementary view raise a fundamental question about Indian thought traditions and perspectives. Are they products of less evolved stages of cognitive development or of higher stages? Are Indians less evolved in their cognitive development or more evolved? When we look around, we find that Indian people are in no way different from Westerners and they do experience polarities as antithetical and, also tend to identify with either of the poles. People certainly want pleasure and do not want pain or sufferings. In this, the average Indian mind is no different from the average Western mind. Even a great warrior and a person of noble character like Arjuna of Mahabharata, in the battle front, wanted to avoid the enormous sorrow and suffering that would entail in waging the war and killing his own people. He weighed that the magnitude of pain would be more than the magnitude of pleasure he will get by winning the war and getting back the kingdom. He was bargaining for lesser pleasure than intense pain. This is what majority of people do anywhere.

The dualist and monist paradigms teaches us that life would be a perennial struggle if we keep on dealing with the opposites as though they are antithetical with an approach–avoidance stance. Purandara Dasa, a famous saint from Karnataka (1484–1564 CE) observes, 'to say I don't want difficulties and want only intense pleasures' is a 'desire for wasted life'. It is simply meaningless to strive for such a life because it is unrealistic. Therefore, we have two alternatives. One is to accept both as inevitable without fighting and without trying to choose one over the other; to treat the so-called negatives of life as opportunities to grow. Such an attitude is inculcated in the minds of people through various ways. One of them is the practice of eating a bit of jaggery made from sugarcane, which is sweet, with little flower buds from *neem* tree, which is bitter, on the Hindu New Year festival Ugādi. In the Bhagavad Gita, Krishna tells Arjuna that a person with equanimity of mind (is one who treats both *sukha* and *dukkha* and *lābha* and *alābha* (profit and loss) as equal. Such a person does not get over-excited at the prospect of pleasure and does not get unduly agitated in pain. This is the complementary view of dualistic paradigm, because it affirms the reality of pleasure and pain.

A step beyond is the monistic, more appropriately non-dualistic (*a-dvaita*), paradigm which urges us to transcend all dualities or polarities including subject-object duality and realize that in reality there is no duality. As per the Advaita Vedānta, as long as our locus of identity is at the realm of subject–object duality, it is inevitable that we experience all kinds of duality and its consequences. Since this is a very abstract and complex issue, the simplest way to explain it is by referring to a discussion of states of consciousness found in the Upanishads.

In Māndukya Upanishad (Nikhilananda, 2000) which has only 12 verses, we find a discussion about transcendental consciousness with reference to the other three states, viz., waking (*jāgrat*), dream (*swapna*) and *sushupti* (deep sleep). It is said that there is a fourth, *turīya,* in addition to the three states. This fourth is an *adhāra* (support, ground, foundation, etc.) for the other three. It is in the background and all our phenomenal experiences of the three states are in the foreground like figure, in the Gestalt psychology sense. So, to know this, to realize this, we have to reverse the figure–ground relationship. Ramana Maharshi tries to make it further simple by comparing this fourth state to a movie screen in a cinema theatre. He observes that whatever movie is projected, the screen by itself supports all of them equally, but remains unaffected.

From this point of view, even waking state and our waking experiences are also like dreams, a virtual reality, having no substance like a holographic display in the sky. Therefore, the 'pleasure–pain' duality has no substantial reality from such a state. To experience that fourth state and being established in it $24 \times 7 \times 365$ is *jīvanmukti*, i.e., liberation of consciousness during one's physical existence. For such a person, polarities and dualities are non-existent and do not affect the way it does ordinary mortals. This requires a radical transformation in awareness, which happens with only a few, one in a million. But all others have to follow a gradual course, if one desires, which entails modification in our *svabhāva* and that can be understood with reference to *triguna*.

Triguna and Positive Psychology

In the beginning, it was mentioned that one way of viewing positive psychology movement from the Indian perspective is to view it as a movement towards *sattva*, from the exclusive preoccupation with *tamas* and *rajas*. These three together constitute *triguna*, which is an important construct in Indian traditions. Though its origins are found in the early Vedic and Upanishadic literature, it is developed in Sāmkhya Darśana (dualistic paradigm) and accepted by all the other *darśana*. Indian thinkers have used the concept of *triguna* to explain everything in the universe, since both matter and mind are viewed as manifestations of *triguna* alone. *Sattva, rajas* and *tamas,* correspondingly represent the principles of illumination and creativity, activity and energy, and inertia. Matter is regarded as preponderantly constituted of *tamas*, the principle of inertia and mind is regarded as preponderantly constituted of *sattva*, the principle of illumination and creativity. Therefore, Indian traditions hold that matter and mind are not polarities as Descartes thought, but a continuum.

Indian worldview holds that these principles operate in various combinations in the entire universe in the structure and function of everything including human beings. Thus, people are differentiated as predominantly *sātvic, rājasic* and *tāmasic* depending on the preponderance of one of the three over the other two. A *sātvic* person is described as one who has discriminative intellect; who is self-controlled, serene, equanimous and steadfast; who is virtuous, generous and gentle; and who is detached and duty-bound without expectations, a seeker of self and aware of the unity underlying all diversities. A *rājasic* person is one who is driven into action by passion, is restless, is struggling; who has more desires, strong likes and dislikes, and pursues sensory pleasures; who is attached to one's social roles; who lacks clear discrimination and has distorted understanding; and who is egotistic. A *tāmasic* person is depressed, lethargic, disinclined to work, negligent, undisciplined, arrogant, hostile, indecisive, ignorant, inadvertent, uncertain and dull. It is to be noted that all the three *guna*s are present in all the individuals, and it is the preponderance of one over the other which leads to the labelling of persons as *sātvic, rājasic* and *tāmasic* type (Murthy and Salagame,[2] 2007).

Henry (2013) summarizes the spiritual approaches very meaningfully as:

> 'Spiritual psychologies advocate living ethically and stress the importance of an interpersonal orientation and the value of quieting the mind. Spiritual practice often encourages people to go within, root out their failings, and develop the capacity to attend to others kindly. There is greater focus in Eastern spiritual practice on acceptance and detachment from desires than seeking and striving after goals found in many Western forms of psychological development. An increasing number of new interventions such as mindfulness, compassion therapy, and reconciliation, are drawn on spiritual practice for inspiration.The marriage of psychological and spiritual approaches shows great promise (2013: 421)'.

All this involve developing more of *sātvic* qualities.

All the Indian perspectives concur that spiritual development involves getting more refined, which means, shedding of *tāmasic* and *rājasic* tendencies and developing more of *sātvic* qualities. Since *sattva* is the principle of illumination transcendental consciousness can be realized only when mind becomes more transparent by losing its opacity (*tamas*). That is why the prayer *asatoma sadgamaya tamasoma jyotirgamaya mrtyorma amrtamgamaya.* All the virtues and character strengths contemporary positive psychology speaks of, are listed as *sātvic* qualities in Indian traditions.

We can compare this movement away from *tamas* and *rajas* towards *sattva*, to a rocket which is launched to outer space. If the rocket has to successfully move upwards, it has to burn its lower stages, which gives an upward thrust and velocity required to escape the gravitational pull of the atmosphere of the earth. Our seer, sages and saints have compared transcendence to such a process where one has to burn the negative tendencies. From Indian perspectives, therefore, positive emotions and individual traits are not the antithetical opposite of negative emotions and traits, but complementary opposites which can be transposed or transformed. Patañjali Yoga Sutra prescribes such a procedure of developing, viz., *maîtri, karuna* and *upeksha* in the face of hatred, downtroddenness and egotism.

Since Sāmkhya and Yoga are recognized as approaches which speak of a gradual evolution of psyche culminating ultimately in *kaivalya*, they teach a way of dealing with opposites in a complementary manner. The Upanishads and the Vedānta being non-dual in perspective teach that the polarities of life be accepted as they are, but not to be identified with. This is what the Bhagavad Gita teaches in *sthitaprajña lakshana*. This is what Krishna says in Verse 57 (Chapter 2), wherein he points out that *sthitaprajña* neither congratulates nor despises anything accepting equally well good and bad and maintaining a sense of detachment with everything.

[2] In the original, it is Murthy and Kumar. In the last few years, I have been using my last name, Salagame.

Implications of Indian Perspectives for Positive Psychology

The implications of Indian perspectives, for positive psychology, may be understood at two levels, viz., theoretical and applied levels.

(1) At a theoretical level, there are many distinctive features of Indian thought which provide a different vision of reality than what is possible from a purely positivist point of view and it is the view that reality is fundamentally spiritual rather than material. From this meta-perspective, there are certain fundamental notions commonly accepted by all *darśana*, which are of ontological and epistemological significance and may be regarded as the core of Indian psychology. Indian traditions and perspectives uphold that '(a) the spiritual reality can be perceived through intuitive faculty; (b) the soul is independent of body; (c) consciousness, *cit*, is different from mind/psyche; (d) body and mind are constituted of the same three principles or guna (*triguna*) and, hence, they are not different in substance; (e) there is life after death; (f) there are paranormal phenomena which can be experienced; and (g) a human being can attain liberation from the cycle of birth and death through self-realization. These and other related views have shaped the way of life in the Indian sub-continent leading to a holistic perspective, in which a human being is understood as biological, psychological and spiritual in nature and is in constant relation with the whole cosmos' (Salagame, 2013b).

(2) At an applied level, the distinctive features delineated above, implies that we need to examine our current lifestyle and revalidate them with respect to what Indian traditions regards as good life or fulfilled life, instead of aping the West mindlessly when Western society itself is turning towards India for spiritual guidance. We have many good things to learn from positive psychology developed in the West, but we need not just limit ourselves to those theories and models. We can develop our own models. In this direction, there are some efforts, which are worth looking into (Anand, Srivastava and Dalal, 2001; Kohli and Dalal, 1998; Misra, 2009; Mohan, Prasad and Rao, 2004; Mohan, Mohan, Roy, Basu and Viranjini, 2004; Naidu and Pande, 1999; Palsane and Lam, 1996; Pande and Naidu, 1992; Rangaswami, 1994; Shinde, 2001; Shinde, 2002; Singh and Misra 2000; Tewari, 2000; Wadhwa and Jain, 1990).

Conclusion

Modern psychology views man from the Darwinian perspective and regards him (her) as the highest of a primate 'man an animal' in simple terms. On the contrary, the spiritual view of India looks at 'man as potentially divine' or as an aspect of the divine (*daivāmsha sambhūta*). These two radically different perspectives on human nature mean a lot as to how we grow and develop ourselves as human beings; how we define ourselves; what type of self-concept we develop; how we lead our lives; what goals we wish to pursue and ultimately how we evaluate our sense of well-being and happiness.

References

Ajaya, S., (1983). *Psychotherapy East and West: A Unifying Paradigm.* Honesdale, Pennsylvania: Himalayan International Institute for Yoga Sciences.

Anand, J., Srivastava, A., and Dalal, A. K. (2001). Where Suffering Ends and Healing Begins. *Psychological Studies*, *48*(3), 114–26.

Baumgardner, S. R., and Crothers, M. K. (2006). *Positive Psychology (1st Ed.)*, New Delhi: Pearson Education.

Diener, E., (1984). Subjective Well-being. *Psychological Bulletin*, *95*(3), 542–75.

Diener E., (2000). Subjective Well-being—The Science of Happiness and a Proposal for the National Index. *American Psychologist*, *55*(1), 34–43.

Diener, E., and Diener, R. B. (2002). Will Money Increase Subjective Well-being? *Social Indicators Research Series*, *57*(2), 119–54.

Diener, E., and Seligman, M. E. P. (2004). Beyond Money: Toward an Economy of Well-being. *Psychological Science in the Public Interest*, *5*(1), 1–31.

Henry, J. (2013). Quieting the Mind and Low Arousal Routes to Happiness. In S. David, I. Boniwell, and A. C. Ayers (Eds.), *The Oxford Handbook of Happiness* (pp. 411–26). London: Oxford University Press.

Hiriyanna, M. (1993). *Outlines of Indian Philosophy.* Delhi: Motilal Banarasidas Publishers Pvt. Ltd.

Kashdan, T. B., Biswas–Diener, R., and King, L. A., (2008). Reconsidering Happiness: The Costs of Distinguishing Between Hedonics and Eudaemonia. *The Journal of Positive Psychology*, *3*(4), 219–43.

Kohli, N., and Dalal, A. K. (1998). Culture as a Factor in Causal Understanding of Illness: A Study of Cancer Patients. *Psychology and Developing Societies*, *10*(2), 115–29.

Kristjansson, K. (2010). Positive Psychology, Happiness, and Virtue: The Troublesome Conceptual Issues. *Review of General Psychology*, *14*(4), 296–10.

Maslow, A. (1971). *The Farther Reaches of Human Nature.* New York: Penguin Books.

Misra, G. (2009). Self and Well-being. *Psychological Studies*, *54*(2), 85–86.

Mohan, K. K., Prasad, V. S., and Rao, P. V. K. (2004). Effectiveness of Spiritually-based Lifestyle Change Programme on Well-being. *Journal of Indian Psychology*, *22*(1), 6–13.

Mohan, Y., Mohan, K. K., Roy, G., Basu, S., and Viranjini, G. (2004). Spiritual Well-being: An Empirical Study with Yogic Perspectives. *Journal of Indian Psychology*, *22*(1), 41–52.

Murthy, P. K., and Salagame, K. K. K. (2007). Concept *Triguna*: A Critical Analysis and Synthesis. *Psychological Studies*, *52*(2), 103–13.

Naidu, R. K., and Pande, N. (1999). *Anāsakti*: The Indian Vision of Potential Human Transcendence Beyond Mechanistic Motivations. In G. Misra (Ed.), *Psychological Perspectives on Stress and Health* (pp. 85–89). New Delhi, India: Concept Publishing Co.

Nikhilananda, S. (2000). *The Māndūkya Upanisad with Gaudapāda Karaka and Shankar's Commentary (Translated from Sanskrit).* Calcutta: Advaita Ashrama.

Ornstein, R. E. (1972). *The Psychology of Consciousness.* New York: Penguin Books.

Palsane, M. N., and Lam, D. J. (1996). Stress and Coping from Traditional Indian and Chinese Perspectives. *Psychology and Developing Societies*, *8*(1), 29–53.

Pande, N., and Naidu, R. K. (1992). Anāsakti and Health: A Study of Non-attachment. *Psychology and Developing Societies*, *4*(1), 89–104.

Peterson, C., and Seligman, M. E. P. (2004). *Character Strengths and Virtues: Handbook of Classification.* New York: Oxford University of Press.

Raju, P. T. (2009). *Philosophical Traditions of India*, (2nd Rev. Ed.), Delhi: Motilal Banarasidas Publishers.

Rangaswami, K. (1994). Self-actualization and Beyond: Union with Universal Self, the Highest Motive from Indian Perspective. *Indian Journal of Clinical Psychology, 21*(2), 45–50.

Ryan, R., and Deci, E. (2001). On Happiness and Human Potentials: A Review of Research on Hedonic and Eudaimonic Well-being. *Annual Review of Psychology, 52*, 141–66.

Ryff, C. D. (1989). Happiness is Everything, or is It? Explorations on the Meaning of Psychological Well-being. *Journal of Personality and Social Psychology, 57*(6), 1069–81.

Salagame. K. K. K (2002a). *An Indian Conception of Well-being.* Paper Presented at the First European Positive Psychology Conference. King Alfred's College, Leicester, UK.

Salagame, K. K. K (2002b). *Some Inputs from India for the Development of Positive Psychology.* Poster Presentation at the First International Summit on Positive Psychology, Gallup Organization, Washington DC: USA.

Salagame, K. K. K. (2003). An Indian Conception of Well-being. In J. Henry (Ed.), *European Positive Psychology Proceedings* (pp.73–80). Leicester, UK: British Psychological Society.

Salagame, K. K. K. (2004). Perspectives on Well-being in the Indian Tradition. *Journal of Indian Psychology, 22*(2), 63–72.

Salagame, K. K. K. (2006a). The Role of Spirituality in Attaining Well-being: Approach of Sanātana Dharma. In A. D. Fave (Ed.), *Dimensions of Well-being. Research and Intervention* (pp. 538–51). Milano, Italy: Franco Angeli.

Salagame, K. K. K. (2006b). *Indian Psychology and Positive Psychology: Birds of the Same Feather Flock Together.* Invited Lecture at the Fifth International Summit on Positive Psychology, Gallup Organization, Washington DC: USA.

Salagame, K. K. K. (2006c). Happiness and Well-being in Indian Tradition. *Psychological Studies, 51*(4), 105–12.

Salagame, K. K. K. (2008a). Systems of Indian Thought–Unit 5. *Self-Instructional Material for M. A. Psychology (Previous)–Block 1–Systems and Contemporary Trends.* Mysore: Karnataka State Open University.

Salagame, K. K. K. (2008b). Indian Thought and Traditions: A Psycho-historical Perspective. In. K. R. Rao, A. C. Paranjpe and A. K. Dalal (Eds.), *Handbook of Indian Psychology* (pp. 19–52). New Delhi: Cambridge University Press of India.

Salagame, K. K. K. (2011). Indian Indigenous Perspectives and Concepts: Developments and Future Possibilities. In G. Misra (Ed.), *Psychology in India Volume IV: Theoretical and Methodological Developments* (pp. 93–172). New Delhi: Pearson Education.

Salagame, K. K. K. (2013a). Well-being from the Hindu/Sanātana Dharma Perspective. In S. David, I. Boniwell, A. Conley Ayers (Eds.), *The Oxford Handbook of Happiness* (pp. 371–83). London: Oxford University Press.

Salagame, K. K. K. (2013b). The Perspectives on Reality in Indian Traditions and Their Implications for Health and Well-being. In A. Morandi and A. N. Narayanan Nambi (Eds.), *An Integrated View of Health and Well-being Bridging Indian and Western Knowledge* (pp. 39–57). New Delhi: Springer India.

Salagame, K. K. K. (2013c). Swāsthya and Stithaprajñata: Health and Well-being from an Indian perspective. In K. Keith (Ed.), *Encyclopedia of Cross-Cultural Psychology* (pp. 1254). New York: Wiley-Blackwell.

Salagame, K. K. K (2014). Positive Psychology in the Light of Indian Traditions. In S. Saini (Ed.). *Positive Psychology in the Light of Indian Traditions.* (pp. 1–21). Haryana, Hisar: Indian Association of Health, Research and Welfare.

Sastri, N. A. (1997). Sramana or Non-Brāhamanical Sects. In *Studies in Jainism.* Kolkata: The Ramakrishna Mission Institute of Culture.

Seligman, M. E. P. (2002). *Authentic Happiness.* New York: The Free Press—A Division of Simon & Schuster.

Seligman, M. E. P., and Csikszentmihalyi, M. (2000). Positive Psychology: An Introduction. *American Psychologist, 55*(1), 5–14.

Sheldon, K. M., Kashdan, T. B., and Steger, M. F. (2011). *Designing Positive Psychology*. New York: Oxford University Press.

Shinde, V. R. (2001). Spiritual and Material Health Associated with Swādhyāyee and Non-Swādhyāyee Families. *Behavioural Scientist, 2*(1), 3–10.

Shinde, V. R. (2002). Emotional Maturity in Swādhyāyee Youths Associated with Divine Brain Trust and Non-Swādhyāyee Youths. *Behavioural Scientist, 3*(2), 81–90.

Singh J. K. and Misra G. (2000). Understanding Contentment in Everyday Life. *Indian Psychological Review, 54* and *55*(4, special millennium issue), 113–24.

Sinha, J. (1985). *Indian Psychology, Vol. II–Emotion and Will.* (Chapter IV, Pleasure and Pain). Delhi: Motilal Banarasidas, (Reprint, 1996).

Snyder, C. R. and Lopez, S. J. (2002). *Handbook of Positive Psychology.* New York: Oxford University.

Tewari, A. K. (2000). Anāsakti and Mental Health. *Psychological Studies, 45*(3), 156–66.

Wachtel, P. L. (1989). *Poverty of Affluence: A Psychological Portrait of the American Way of Life*, New York: New Society Publishers.

Wadhwa, B. S., and Jain, R. (1990). Attitudes of Teachers Towards Four Goals of Life of Hindu Morality—Dharma, Artha, Kāma and Moksha. *Journal of Psychological Researches, 34*(1), 6–9.

Wilber, K., Engler, J., and Brown, D. P. (1986). *Transformation of Consciousness: Classical and Contemplative Perspectives.* Boston, MA: Shamabala.

CHAPTER 2

Progress in Positive Psychology: Some Reflections from India

Jitendra K. Singh

Defence Institute of Psychological Research, Delhi

Abstract: The foundation of positive psychology as one of the core sub-disciplines of psychology is rooted in identifying and enhancing positive states, positive traits, positive emotions and positive relationships of the human beings. It assumes that these positive aspects of a human being facilitate his/her strengths and, hence, prevent all the behavioural aberrations which overpower his/her thinking, feeling and action. Since its inception during the late 1990s, the field has grown tremendously and has attracted attention of researchers across the world. The six virtues and twenty-four-character strengths are considered to be ubiquitous and are claimed to be endorsed by every culture. However, the conceptualization of human psychological strength within the rubric of positive psychology subscribes to the Western idea of individual. On the contrary, the religio-philosophical traditions of India, viz., Hinduism, Buddhism and Jainism deal comprehensively with various positive aspects of the individual. These traditions offer substantial insight about the conceptualization of human strength in terms of attributes, emotions and relations; a practical philosophy to deal with daily life issues; and a framework of healthy human development. Against this background, the chapter at first tries to highlight core positive aspects and constituents in addition to virtue and character strength. Thereafter, it discusses in detail, some of the key Indian concepts dealing with positive aspects of human being such as *anasakti, ahamkara, santosh, triguna* and wisdom and its relevance to describe human strength in the Indian context.

Conceptual Root and Contemporary Development

The foundation of psychology as a discipline of human behaviour in general and a helping profession in particular, is rooted in looking into multiple facets of behaviour including behavioural aberrations and subsequently designing intervention programme to make the individual capable of adjusting with the demands of the social world. Such an approach of helping profession, perhaps, takes conceptual and methodological insights from the clinical model of the human behaviour and has flourished across the world. Over the years, assessment, diagnosis and management of behavioural disorders have become the core areas and have been able to deliver services to different sectors of the society. The predominance of disease model in intellectual thinking has made the discipline, largely, a science about healing.

The beginning of the 21st century has witnessed the emergence of a new field that focuses primarily on the positive aspects of human life in the intrapersonal, interpersonal, societal and organizational domains. The field is popularly known as positive psychology. It focuses on identifying and enhancing positive states, positive traits, positive emotions and positive relationships and assumes that identifying and enhancing these positive aspects facilitates growth and development in healthy direction.

In order to evolve a holistic understanding of the human being, the field of positive psychology classifies its role at three levels (Seligman and Csikzentmihalyi, 2000). At the *subjective level* it deals with valued subjective experiences such as well-being, contentment and satisfaction (in the past), flow and happiness (in the present), and hope and optimism (for the future). At the *individual level*, it is about positive individual traits such as love, courage, interpersonal skills, aesthetic sensibility, perseverance, originality, forgiveness, future mindedness, spirituality, high talent and wisdom. At the *group level*, it focuses on civic virtues and institutions that take individuals towards better citizenship, responsibility, nurturance, altruism, civility, moderation and work ethics. Considering the saliency of all the three levels in human life, positive psychology has been defined as a science of positive subjective experiences, positive individual traits and positive institutions to improve the quality of life and prevent pathologies that arise when life is barren and meaningless (ibid.).

Though positive psychology has recently come into limelight, its roots can be traced to the 1930s. As highlighted by Peterson and Seligman (2004), the root of the field can be traced in Terman's (1939) study of giftedness and marital happiness (Terman, Buttenweiser, Ferguson, Johnson and Wilson, 1938), Watson's (1928) writing on effective parenting and Jung's (1933) work on search and discovery of meaning of life. In later years, the influence of positive psychology has been reflected in the humanistic approach of explaining personality and behaviour. However, the birth of positive psychology as a full grown sub-discipline during the latter half of 1990s and beginning of the 21st century can be seen in the pioneering contributions of a number of scholars. Some of the concepts which received attention were *positive experience* (Kahneman, 1999), *subjective well-being* (Diener, 2000), *optimal experience* (Massimini and Delle Fave, 2000), *optimism* (Peterson, 2000), *happiness* (Myers, 2000), *self-determination* (Ryan and Deci, 2000) and *relationship between positive emotions and physical health* (Taylor, Kemeny, Reed, Bower and Gruenwald, 2000).

In the field of positive personal traits, Deiner (2000) worked on subjective well-being. The study on subjective well-being focuses on how people cognitively assess their existence. He found a link between macro-social conditions and happiness. He elaborated, how a person's values and goals mediate between external events and the quality of happiness. Peterson (2000) did valuable work on optimism and dealt with cognitive, emotional and motivational components of optimism, its mechanism and when it begins to distort the reality. Myers (2000) discussed the factors that promote happiness, religious faith, economic growth and close personal relationships. Ryan and Deci (2000) studied self-determination. They claimed that when three needs, i.e., relatedness, competence and autonomy, are satisfied, it results in personal well-being and social development. The individual is intrinsically motivated and seeks greater challenge to fulfil his potential. They have also described the social contexts that support or hinder it. Schwartz (2000) indicated that excess of freedom leads to a greater burden of responsibility which can give rise to depression, dissatisfaction, feelings of insecurity and regret. According to him, cultural values and norms are important for a meaningful and satisfactory life. Vaillant (2000), from his lifespan longitudinal research, summarized the contributions of mature defenses like altruism, sublimation, suppression, humour and anticipation for a joyful life. He emphasized that proactive and creative solutions are necessary to break the mold of victimology.

In order to position itself as an equally relevant and important sub-discipline, there has been significant amount of publications in this field for the last one decade and a half. These publications are mainly in the form of books and research articles. These publications have greatly helped the new scholars to take valuable insights for furthering research in this area. Some of these publications are—*The Handbook of Positive Psychology* (Snyder and Lopez, 2002), *Authentic Happiness* (Seligman, 2002), *Psychology of Human Strengths* (Aspinwall and Staudinger, 2003), *Flourishing* (Keyes and Haidt, 2003), *Positive Psychology Assessment: A Handbook of Models and Measures* (Lopez and Snyder, 2004), *Positive Psychology in Practice* (Linley and Joseph, 2004) and *Handbook of Methods in Positive Psychology* (Ong and Van Dulmen, 2007) and *Encyclopedia of Positive Psychology* (Lopez, 2009). One of the most notable publications is *Character Strengths and Virtues: A Handbook and Classification* (Peterson and Seligman, 2004). It is similar to Diagnostic and Statistical Manual (DSM) published by American Psychiatric Association. Contrary to psychological disorders, the handbook describes strengths and virtues. Each chapter in the handbook of character strengths and virtues is about what is known about strengths, their definitions, paradigm, historical and cross cultural background, measurement, correlations and effects of having or lacking a particular strength, gender differences and interventions. Based on the studies on Character Strength and Virtues (CSVs), Peterson and Seligman (2004) identified six virtues and 24 character strengths that are claimed to be ubiquitous (Park, Peterson and Seligman, 2006).

During its short journey, the field has covered a great deal of conceptual and empirical plains and has incorporated a number of concepts in its fold other than virtues and character strengths. These concepts are briefly discussed in the subsequent paragraphs.

Positive Experience

Experiences are one's interaction with immediate environment comprising objects, stimuli and persons. It shapes an individual's emotions, positive as well as negative. Positive emotions are essential to lead a good life. Positive experiences buffer the effect of negative events and emotions and give a sense of control and mastery in everyday life. People who experience more positive events report greater well-being (Strand, Reich and Zautra, 2009). Csikszentmihalyi (1990) worked on positive experiences and developed flow theory. According to him, flow refers to a state of optimal experience with total absorption in the task at hand. Flow is a characteristic of good life. It develops positive feelings when one's experiences become focused and other concerns are eliminated. Positive feelings result from this total focus. Flow can be understood in terms of external facilitators such as one's location of activities and internal facilitators like concentration, happiness and attention to experiences.

Diener (2000) reported that environmental events and a person's cognitive and emotional reaction to them are central to one's good health and well-being. Studies have reported that positive emotions are linked to happiness and contentment, but not related to lowering down the intensity of negative emotions (Diener, Oishi and Lucas, 2009). However, it has been found that positive experiences have 'buffering effects' which leads to lessening the impact of negative experiences. Buffering effects of positive experiences are possible under high stress (Zautra, Potter and Reich, 1997). Positive experiences lead one to pleasurable engagement with environment and create positive feelings. It is connected with survival against odds and facilitates motivational, physiological and cognitive processes. Positive experience in the context of cognitive strategies helps to frame situation in more positive light. It expands one's capacity to organize ideas and creative problem-solving strategies.

In addition, positive experiences help to bring people together and reinforce a bonding among them. This bonding provides a positive feeling which is central to lead a good life. So, positive experiences provide good feeling about life by resulting in personal mastery, control of the life events, and finally, opportunities for growth and well-being.

Positive Emotions

Positive emotions are essence of a good life. They are brief experiences of feeling good in present and increasing chances that one will feel good in the future (Froh, Kashdan, Ozimkowski and Miller, 2009). The study of positive emotions is a recent development in the field of positive psychology. One of the initial studies in this area was conducted by Isen (1987) which dealt with the role of positive emotions in cognitive processes. It showed that when people experienced positive emotions, they demonstrated broadening of thinking by associating more unusual words with neutral words, flexible thinking and creative problem solving. Another study suggested that positive emotions and broad-minded coping have positive relationship (Fredrickson and Joiner, 2002). Thus, positive emotions make one feel good in present and create chances of feeling good in future (Fredrickson, 2002). It was also reported that gratitude appeared to be one of the most commonly experienced emotions at the time of terrorist attacks on September 11, 2001 (Fredrickson, Tugade, Waugh and Larkin, 2003). The archival data of newspaper account showed children to be thankful after 9/11. It revealed that gratitude play an important role in coping with adversity. Thus, positive emotions function to broaden one's thinking pattern as well as contribute to well-being.

From an evolutionary perspective, people tend to give more attention to negative event than positive ones. We tend to form bad impressions and stereotypes more quickly than good ones. This is because negative events tend to have lasting negative impact, which is not found in the case of positive events. However, the idea here is to use positive events such that they can have lasting good/positive effects. Positive emotions can predict future resilience and improved coping by broadening one's cognitive and behavioural domains. Fredrickson (2002) proposed that positive emotions can broaden thought–action repertoires that later build enduring physical, intellectual and social resources. Physical skill/resources are acquired in play. They aid in one's survival. Intellectual resources can be developed by experiencing positive emotions like joy and happiness. Joy motivates a person to explore environment and gather greater knowledge about it. This helps to acquire basic knowledge (e.g., to learn who is in the environment) and applied knowledge (e.g., to learn who is approachable enough for help in future).

Social resources can be developed by learning to express gratitude. It is a way to replay kindness. Thus, repaying of kindness results in strengthening relationships and building social capital (Fredrickson, 2001). Also, this has lasting impact on one's social relationships. A study in this area also reported that happy individuals live longer, earn more and enjoy loving relationship (Lyubomirsky, King and Diener, 2005).

Positive Ethics

Positive ethics shifts the focus from wrongdoing to more balanced and integrative approach that includes aspiring for the highest ethical potential (Handalsman, Knapp and Gottlieb, 2002). It guides individuals to be morally correct in their conduct. Ethics represent a dynamic enterprise that allows psychologists to respond better to changing conditions and situations. They may include idea of what is good, valuable and best of tradition. Ethics are often understood from the point of obedience to rules, but positive ethics mean striving for highest level of ethical excellence. It urges individuals to live up to high ideals such as (1) to integrate personal values and professional ethics; and (2) to encourage individuals to fulfil their highest personal idea through their career.

Emphasis on positive ethics allows drawing upon philosophical and scientific perspectives for judging an action and choosing the best way to promote ethical ideals. Apart from this, positive ethics provides the psychologist with a framework that gives meaning to their work. Hence, the purpose of positive ethics is to relook all aspects of professionals and scientific aspects form ethical perspective. It has implications for creating a workforce which values ethics and morality in all walks of life, thereby creating responsible and morally upright individuals.

Empathy

Empathy is the ability to understand events/situation from other persons' perspective, so as to acquire insight into their psychological world. Empathy includes two distinct activities. First, it includes imagining what other person is thinking or feeling as a result of a situation, considering the values and goals of that person. The second point involves visualizing oneself in others' situation (Stocks and Lishner, 2009). Empathy involves listening to the other person in order to become aware of his/her internal states. It is a feeling about what another person is feeling such as experiencing anger in the context of the other person's suffering or unjust treatment.

Studies have shown that empathy results in pro-social behaviour and is inconsistent with anti-social behaviour. It is also related to conflict management, improves communication among partners, relationship satisfaction and consideration towards partner's needs (Stocks and Lishner, 2009). On the other hand, the way attribution is made regarding other person's behaviour, affects empathetic feelings. An individual tends to explain his/her own behaviour as a consequence of situation and explain behaviour of others as a consequence of personality traits, particularly in case of negative event. However, in case of empathy, the observer makes situational attribution for others' behaviour.

Thus, empathy augments the value of the other person and reduces the likelihood of blaming and derogating the victim. In addition, empathy shapes the attitudes towards stigmatized groups and social causes. It reduces prejudice and discrimination against racial minority groups. Hence, empathy is a potential force to improve relationships and attitude towards environment.

Hope

Hope is a positive mental state that enables one to achieve goals in the future (Edwards, 2009). It also indicates the expectation of an individual to attain a goal. It is also described as an emotion that is guided by cognitions and influenced by environmental conditions (Averill, Catlin and Chon, 1990). Hope is also defined as 'a positive motivational state that is based on an interactively derived sense of successful (1) agency (i.e., goal-directed energy), and (2) pathways (planning to nut goals)' (Snyder, Irving, and Anderson, 1991, p. 287). Snyder (2000a, 2000b) conceptualized hope as having cognitive and motivational components which jointly facilitate goal-setting behaviour, its pathways and finally goal-directed thinking. For hope, goals need to be sufficiently important to individuals and with probability of attainment component so that individuals can imagine themselves being able to reach goals. Another component of hope is will power that motivates individuals to begin and sustain their efforts toward the goals.

Studies on hope have reported that it is related to well-being and several positive outcomes of life such as psychological adjustment, physical health, academic and athletic performance. It is linked to coping behaviours among individual surviving illness and health outcomes such as perceived health and sense of vigour. Hope is related to positive outcomes in adults like better performance on standardized achievement tests and belief about ability to achieve goals. It has been reported that hope is linked to improved mood and treatment outcome. Thereby, it becomes evident that hope is a useful construct in life.

Optimism

Optimism refers to an explanatory style that attributes positive events to personal, permanent and pervasive causes, and negative events to external, temporary and situation-specific ones. In contrast, pessimism explains positive events through external, temporary and situation-specific attributions and negative events through internal, permanent and pervasive ones (Seligman, 1991, 2002). Optimism is generally described in terms of expectancies from self and others.

People with optimism expect good things to happen. Their ways of problem solving and coping differ from pessimistic individuals who expect bad thing to happen. This orientation has implications for psychological and physical well-being.

Studies on optimism have generally viewed optimism and pessimism as bipolar dimension. It has been reported that optimistic individuals experience less distress at the time of adversity than pessimists. Optimism was found to be associated with beneficial changes in distress over time. It has been found that optimistic individuals are confident to attain a goal, continue to try, whereas pessimistic individuals are doubtful and try to escape by using temporary distractions or even stop trying. Optimists have greater psychological well-being due to better coping strategies. They report fewer physiological symptoms at the time of distress and more adaptive immune functioning than do pessimists. They engage in problem-focused coping and proactive process to promote well-being. They make effort to reduce risk and safeguard their health. Therefore, such individuals make themselves less vulnerable to health issues. However, too much optimism is damaging because it might lead people to ignore potential threat or overestimate their capability to deal with it resulting in poor outcomes. Thus, an individual can obtain better outcomes by maintaining a positive and realistic view of situations. In contrast, pessimists tend to give up in face of serious problems and may indulge in substance abuse, excessive alcohol use as form of escapism. In some cases, people give up not just on problems they face, but also their lives, by committing suicide. Hence, pessimism is even stronger indicator of suicide than depression. These coping tendencies used by pessimistic individuals can obstruct their psychological well-being.

Psychological Capital

Psychological capital refers to the ability of a person to allocate attention so as to generate positive experiences in the present, and in ways that are likely to provide positive experiences in the future as well (Csikszentmihalyi and Nakamura, 2009). Thus, the ability to choose desirable focus of attention and keep one's attention on desired objects become the psychological capital. The concept of psychological capital is based on research on flow which is a subjective state of involvement in any activity. Some individuals are able to find such subjective state (flow) and enjoy even in difficult situations of life. The presence of this trait in individuals came to be called autotelic personality and the same was extended into a lifespan concept called, 'psychological capital (PsyCap)'. PsyCap is based on the premise that one's quality of life is determined by where a person chooses to invest his/her attentional resources (Luthans, Youssef and Avolio, 2007).

Mindfulness

Mindfulness is an alert state of mind characterized by noticing new things. It is a feeling of engagement or involvement and awareness of what is there in present. In contrast, mindlessness, is an inactive state of mind characterized by reliance on past, outdated categories, lack of awareness of one's social and physical world. Mindfulness results in better learning, more intelligent and creative products such as positive effect. Also, people are more attracted to individuals who are mindful. Mindful learning is similar to learning probable truth as opposed to mindless acceptance of statements. Learning mindfully results in improved creativity, high attention level and improved memory (Langer, 2009a and 2009b). To become mindful, one needs to learn to switch models of thinking about ourselves and the world. In addition, meditation is one way to attain mindfulness. The idea of mindfulness has importance for one's health and well-being.

Self-Efficacy

Self-efficacy refers to belief in one's ability to produce desired outcomes. It determines how much effort one wants to put in to overcome obstacles. It focuses more on what one will do particularly under changing and challenging situations. It is not a motive to control, but conviction in one's ability to perform effectively. It is defined as 'one's conviction (or confidence) about his/her abilities to mobilize the motivation, cognitive resources and courses of action needed to successfully execute a specific task within a given context' (Stajkovic and Luthans, 1988: 66).

The development of self-efficacy beliefs in an individual is influenced by two factors. First, the development of symbolic thought and capacity to see cause and effect relationship facilitates self-reflection. This creates an understanding of how one's actions affect environment and increases one's self-awareness. Second, family and environment play a major role in facilitating or inhibiting the development of strong self-efficacy by responding to child's actions. Self-efficacy beliefs shape throughout life as people integrate information from different sources. It is influenced by performance experiences and one's attempt to control a situation.

Influence of self-efficacy beliefs has important consequences for psychological functioning especially in the domains of psychological adjustment, physical health, self-regulation and psychotherapy.

Psychological Adjustment

A person who has a sense of control over environment, behaviour and thought experiences happiness and well-being. On the other hand, lack of such control may lead to psychological problems like depression and inferiority. Low self-efficacy in threatening situation may lead to dysfunctional anxiety and avoidant behaviour, problem of substance abuse and eating disorders.

Physical Health

Self-efficacy encourages one to adopt healthy behaviour and facilitate behavioural change in the face of difficulty. It influences biological processes and effects the body's physiological responses to stress and immune system.

Self-regulation

Self-efficacy influences self-regulation of one's behaviour. It impacts how an individual sets goals. It also influences people's choice of goal-directed activities, expenditure and persistence in the face of challenge. This produces desired results and strengthens self-efficacy that also facilitates problem solving and decision making. People who have confidence in their ability to solve problems can effectively use their cognitive resources and arrive at a better solution and greater achievement than those who doubt their abilities.

Psychotherapy

Psychological intervention may be effective in enhancing self-efficacy. It can be done by identifying experiences that strengthen behaviour which contributes to self-efficacy. The strategies to enhance self-efficacy are based on five sources of self-efficacy. These are: performance experience, vacations experience, imagined experience, verbal persuasion, and physiological as well as emotional states.

Collective Efficacy

The notion of collective self-efficacy is based on the idea that in order to accomplish goals in groups, organizations look for ability to identify the strengths of other persons. A group that has confidence in its abilities to accomplish goals is more satisfied. The collective efficacy beliefs of a group or team can be increased or reduced by commenting on their abilities. In case of businesses, organizations and government success depends on the ability to coordinate efforts to achieve goals or resolve conflicts. An individual's happiness is also dependent on his/her ability to cooperate and negotiate with other people. So, in the present scenario of collaboration between commerce and government, the notion of collective self-efficacy becomes more important.

Resilience

Resilience refers to one's ability to positively adapt to adversity and recover from challenges. It involves functioning well in times of adversity, bouncing back from disturbing experience and achieving new level of positive or normal adaptation when conditions improve (Masten, Cutuli, Herbers, and Reed, 2009). It is also defined as 'the capacity to rebound or bounce back from adversity, conflict, failure or even positive events, progress and increased responsibility' (Luthans, 2002a: 702).

Studies on resilience began with studying people who were at risk of developing problems, e.g., children at risk due to hazardous rearing condition. Resilience research was facilitated by emergence of developmental psychopathology. The findings of the studies revealed that resilience was the key factor in preventing problem and leading to positive adaptation (Masten and Reed, 2002). Resilience develops from many resources and processes that lead to positive outcome. This is because the pattern of behaviour results from many interactions within the person and between person and environment.

The notion of 'positive adaptation' is based on two broad domains called *psychopathology* and *competence*. Positive adaptation from the perspective of mental health is viewed as developing effective coping skills for healthy adjustment and well-being of the individual. At the same time, it also ensures the absence of any clinical disorders or symptoms in the individual. Positive adaptation from the perspective of mental health is viewed as developing effective coping skills for healthy adjustment and well-being of the individual. At the same time it also ensures the absence of any clinical disorders or symptoms in the individual.

Developmental scientists define adaptation in terms of competence in developmental tasks (Cutuli and Masten, 2009). As per developmental psychology, achieving competence in one development stage facilitates the development of later competencies. One stage instills the fundamental skills on which future competence is developed.

Threats to positive adaptation may result from individual's attributes or situations. Researchers have found certain factors that lead to positive outcomes in face of risk or adversity. These factors are called *promotive* and *protective* factors. Promotive factors refer to good outcome in general regardless of presence of risk. Protective factors result in moderating risks when adversity is high. It has been found that most common protective and promotive factors include individual attributes and their relationships such as secure attachment. Close relationships have been found to moderate risk across all age groups. Such relations provide warmth, security and reassurance in the face of adversity. One key protective factor is one's social connections with people other than family such as teachers, mentors and peer relations. These relations provide certainty to one's life if one is living through adversities and they provide life-long social support throughout life. In addition, people with good problem-solving skills and regulation of emotions, attention and actions are better able to adapt to the adversity. So, prevention practices can be designed to set positive goals and strategies to increase resilience and protective factors.

Virtue

The term *virtue* is derived from Greek word 'arte', which means *excellence*. Its meaning is rooted in cultural and religious traditions of the world such as Hinduism, Christianity, Judaism and Buddhism. It describes the moral strength that enables a person to lead an ethically ideal life. A virtuous individual is attracted to knowledge, good relationships and will act in a manner to reach the desired goals. Thus, the essence of virtue is to attain harmony between duty and desire. Socrates identified four virtues: courage, justice, temperance and intellectual resourcefulness. Aristotle expanded this list to include virtues such as liberty, pride, honesty, wit and friendship. He emphasized the use of practical intellectual resourcefulness in enacting these virtues. Aquinas (*ca.* 1265 CE/1966) added three theological virtues—faith, hope and clarity to Socrates' list of virtues. Confucian virtues include reverence, love within the family, righteousness, honesty, benevolence and loyalty to the state. All the above-mentioned categorization of virtue presents diverse perspectives as far as the saliency of virtue in life is concerned. However, none of them can be completely accepted because such perspective varies with time, situation and culture of its proponents. So, enactment of virtues varies with the worldview and is mostly situation dependent.

Positive Psychology in India: Some Reflections

It is generally observed that the conceptually and empirically driven psychological theories and models developed in Western cultures often become a source of inspiration for furthering research in the relevant areas in non-Western cultures. The studies carried out in non-Western cultures with the basic premise of universal acceptability or non-acceptability finally establish the extent of universality of the particular theory or model across cultures. Perhaps this is one of the dominant trends of research in the 21st century psychology (Singh, 2009). The development in the field of positive psychology in non-Western part of the world including India follows this trend. In recent years, positive psychology has become one of the core areas of enquiry being observed in the sudden shift in the interest of scholars. The manifestation of which is being witnessed in the form of empirical studies, conferences, associations, journals and magazines in this area.

India being the cradle of one of the rich intellectual traditions of ancient civilizations along with Babylonia, Egypt, China, Greece and Rome (*see* Mayer, Lin and Korogodsky, 2011; Millon, 2012) offers a suitable vantage point to discuss the three levels outlined by Seligman and Csikzentmihalyi (2000) in order to develop a holistic understanding of the human being. In recent years, the revivalist tradition being advocated within the rubric of Indian psychology has made significant developments in this direction (*see* Cornelissen, Misra and Varma, 2011; Misra, 2011, Rao, Paranjpe and Dalal, 2008; Salagame, 2011). Some of the key concepts which deal with the holistic understanding of human beings are discussed in this section. These concepts are indigenous in origin and offer a relatively better conceptualization of human being in terms of strength, well-being and happiness.

The ancient texts such as Vedas, Upanishads, Bhagavad Gita, Charaka Samhita, Mahabharata, and *Kamasutra* often mention terms like *prakriti, jiva, purusa, svabhāva, atman, ksetrjna* and *pudgala,* for describing a person (Dwivedi, 2002). *Advaita*, a sub-school of the Vedanta system of Indian philosophy mentions the term *jiva* for the term 'person' (Paranjpe and Rao, 2008). In the literal sense, the term *jiva* is used to characterize any form of living being. *Jiva* is conceptualized as a multi-layered entity constituting of five layers nested in one another. The outermost layer of the *jiva* is called *annamaya kosa* or 'sustained by food' and designates body. The second layer is called *pranamaya kosa* or 'sheath of the vital breath'. It activates breathing and other bodily processes necessary to keep the organs functional. The third layer is *manmaya kosa* or 'mental sheath' which includes the sense organs and their functions, as well as various processes collectively called 'mind'. The fourth layer is *vijnanamaya kosa* or 'cognitive sheath' and includes intellect involving ideas, concepts or constructs used to explore the world. The fifth and the inner-most core of the *jiva* is called *anandamaya kosa* or 'joyous sheath' or the sheath of bliss. This layer is called 'the seat of joy' because it reflects bliss (*ananda pratibimba*) which is characteristic of the true self that is *Atman*, which is claimed to be identical with the *Brahman*, the core of ultimate reality. A person can reach this highest stage of evolution through continuous spiritual practice and experience a complete alteration in the sense of identity. In such a stage, the experience of bliss and the experiencer no longer remain separate. In essence, the *Advaita Vedanta* conceptualization of person is perhaps one of the earliest descriptions of the whole person in terms of composition of five integrated layers.

The neo-vedantic tradition pioneered by Sri Aurobindo has also developed a framework rooted in consciousness and has articulated a structure of the being. Sri Aurobindo has advanced two systems, one is concentric and the other is vertical. The concentric system consists of the outer or surface being, the inner being, and supporting both of these, the innermost being or the psyche. The first two have three parts: mental, vital and physical. The vertical system consists of various levels or gradations of consciousness below and above ordinary consciousness. These levels of consciousness constitute the inconscient, the subconscient, the physical, the vital, mind, higher mind, illumined mind, intuition, overmind, super mind and *sacchidanand* (Dalal, 2001; Varma, 2011).

Dwivedi (2002) has discussed the concept of *purusa* as mentioned by Charaka the great physician of ancient India. Charaka discussed three types of *purusa,* viz., *sadadhatuja purusa*, *cetanadhatuja purusa* and *caturvimsatitattvatmaka rasi purusa*. Among these three categories of *purusa* he emphasized on *sadadhatuja purusa*. According to Charaka, *purusa* is an assemblage of *sadadhatus* (six elements) which is condensed form of 24 general elements of *purusa*. Being *avyakta* (the premier substance), *ksetrajna* (the knower), *sasvata* (the eternal), *vibhu* (glorious) and *avyaya* (imperishable) are the characteristics of *purusa*. Charaka described that the *purusa* evolves from the *avyakta*; from the *avyakta, buddhi* (intellect) emerges which gives rise to *ahamkara* (the selfhood) and, thereafter, from *ahamkara* five *mahabhutas* evolve. According to Charaka, the five basic elements (e.g., ether, air, fire, water and earth), *buddhi* (intellect), *avyakta* and *ahamkara* constitute *mahabhutas* or the grand elements. In addition to these, *mahabhutas* are also characterized by 16 *vikaras* which constitute five *budhindriyas* (sense organs), five *karmendriyas* (motor organs) and five elements along with the *manas*. Buddhism and Jainism which form part of the ancient Indian religio-philosophical tradition also deal with conceptualization of human being. Buddhism discusses an aggregate model and a network model in this regard. The aggregate model offers a holistic conceptualization of a person. It proposes that all the experiences are analysed and categorized in five aggregates (*skandha*) such as form (*rupa*), feeling (*vedana*), perception (*samjna*), formation (*sanskara*) and consciousness (*vijnana*). The network model discusses main-mind (citta) and mental episodes (caitta) which determine cognitive and behavioural dynamics (Ananda and Prasad, 2011; Duerlinger, 2008).

In Jainism the term *jiva* (self) stands for a living organism, a biological being, a conjoint psycho-somatic, psycho-physical and conscious entity (Jain, 2008). *Jiva* is conceptualized as knower, enjoyer or experiencer and active agent. The characteristics of *jiva* as discussed in Kundakunda's *Panchastiykayasara* are: (1) it has bio-energies or external manifestations of life; (2) consciousness is the essence and internal life source of *jiva*; (3) *upayoga* (i.e., conscious attentiveness, psychic exertion, function or manifestation of consciousness) is the characteristic of *jiva*, which distinguishes living from non-living; (4) the self is the lord or architect of his own destiny (*prabhu*), who through own efforts obtains full freedom bearing full moral responsibility for conduct; (5) the self is an active agent (*karta*) contrary to the concept of passive spectator *purusa*, as developed in *Samkhya* system; (6) since *jiva* is *karta* and *bhokta* it is responsible for its own action; and (7) *jiva* is of the same dimension as the body in which it resides, with the result that sensation (*samvedana*) is felt in all parts of the body (ibid.).

Indigenous Constructs Dealing with Positive Aspects of Human Being

In addition to conceptualization of human being in the religio-philosophical traditions of India, there are indigenous constructs which have been empirically examined in Indian cultural context. These constructs in some way or the other also deals with positive aspects of human beings. These constructs are briefly discussed in subsequent paragraphs.

Anasakti

Anasakti, in the Indian philosophical thought, is defined as intense activity performed without much concern about success or failure which results in task excellence through unification of the actor with the act (Pande and Naidu, 1992). Pande and Naidu (1992) delineated some of the core characteristics to conceptualize *anasakti*. These were effort orientation, emotional equipoise while confronting success or failure, low concern for obtaining extrinsic rewards, and intense effort to achieve excellence. They developed a scale to assess *anasakti* based on the construct of *Sthitaprajna* (man of steady wisdom) described in the Gita. This scale taps features such as outcome vulnerability, attachment, effort orientation, endurance and equipoise and physical sensual non-identification. They reported that *anasakti* (a dispositional attribute) was a more powerful predictor of strain scores than stress.

Ahamkara

The term *ahamkara,* in Indian context, is often translated as ego, egotism or egoism. The term ego is more close to the term *abhimana,* or *garva* than *ahamkara*. On the other hand, *ahamkara* refers to the sense of doer-ship and ownership which represents cognition and feeling associated with 'me' and 'mine'. The Vedanta tradition uses *ahamkara* along with *buddhi*, *citta* and *manas* as the aspects of *antahkarana catustya*, the quartlet of the internal organ. *Ahamkara* also denotes a feeling of individuality or uniqueness, one's identification and the sense of differentiation of oneself from the other, the 'I' from 'not-I'. Thus, *ahamkara* refers to the subject in subject–object duality in the realm of mind/psyche and, hence, nearer to the term 'self-sense' (Salagame, 2011). The empirical studies on *ahamkara* outline its four components, viz., identification, individuality, agency and separation (Salagame *et al.*, 2005). The concept represents a meta-construct which embraces many of the concepts associated with self and identity such as locus of control, self-efficacy, self-esteem, individuality, relational self, individualism–collectivism, ego-boundary and autonomy (Salagame, 2011).

Santosh

In Indian thought, limiting the needs is frequently considered as the most valued manageable affair to make sustainable progress in life. Liberation *(moksha)* as the ultimate goal of life is attained in this life *(jivanmukta)* by following the path of *karma* and *dharma* in everyday life. The notion of *karma* in this regard, stands for dedication towards work without excessive concern with rewards. The concept of *dharma* emphasizes limited assets (*aparigrah*) sufficient enough to sustain one's life.

Within this perspective the notion of *santosh* (translated as contentment) is approached from an attitude of restraint as well as the experience of need satisfaction. It is also a subjective experiential state of lack of dissatisfaction where the person does not expect much from others but feels contented and happy with his present conditions. The experience of subjective wholesomeness (*santosh*) leads to real pleasure, happiness and satisfaction, and endows the person with inner strength and power to strive for progress and development. Extending this conceptualization the notion of *santosha* (contentment) in everyday discourse was explored by Singh and Misra (2000) using young adults, older people and saints as participants. The connotations of contentment were somewhat different for the common men as compared to the saints. The latter shared a rational worldview while viewing contentment in different facets of life.

Triguna

Triguna which characterizes mental attributes of the person, viz., *sattva* (goodness, harmony), *rajas* (passion, mobility, energy) and *tamas* (dullness and mass/inertia) has been one of the most popular constructs. Among these, *sattva* deals with positive aspects of human being. The *Sankhya* system states that increasing *sattvic* characteristics from *tamas* to *rajas* is important for spiritual evolution. Purification of mind and lessening of *tamas* are considered achievable goals through methods such as *tapas* (penance), *jnana* (intuitive knowledge), *brhamacharya* (restraining and turning the sense inward) and *shraddha* (dedication and devotion to realize the self). Over the years *triguna* has received much attention from scholars. A number of studies have been carried out to conceptualize it and to develop measures to assess *triguna* (*see* Murthy and Salagame, 2007). It is now empirically established that *triguna* is an indigenous personality construct and is more relevant to understand personality in Indian context (Singh and Misra, 2013; Singh, Misra and DeRaad, 2013).

Wisdom

Attempt has been made to conceptualize wisdom in Indian context using its description in Bhagavad Gita as well as layman understanding of it in contemporary Indian society (Jeste and Vahia, 2008). The study has identified 10 domains of wisdom, viz., knowledge of life, emotional regulation, control over desires, decisiveness, love of god, duty and work, self-contentedness, yoga or integration of personality, compassion/sacrifice and insight/humility. Srivastava and Misra (2001) reported that in the Indian culture, the meaning of intelligence as shared by people constitutes not only cognitive competence, but also social, emotional and competence in action. The Indian notions about the construct of emotional intelligence were examined by Sibia and Srivastava (2003), and it was found that emotional intelligence in India is understood in terms of being high on pro-social values, social sensitivity, action tendencies such as discipline, persistence, practical and responsible as well as affective states such as happiness and optimism.

Srivastava and Misra (2003) conducted an analysis of *suktis* in Sanskrit related to happiness and wisdom. They reported that in India happiness and wisdom can be organized around the following themes.

Contentment

Renunciation of craving for what is not obtained and being satisfied with what comes unsought.

Control over Desires

This included interrelated concepts such as *jitendriya* (control over impulses, emotions and actions). *Jitendriya* literally means one who has conquered the senses. *Chittaavriti nirodha* (regulation of mental activities) is considered of paramount importance in yoga.

Surrender

An attitude of surrender to /union with the divine will is seen as a means of ending suffering.

Non-attachment

Focus on working constantly without being attached to outcomes and reserving the power of detaching from everything. Equanimity in opposites—maintaining composure across circumstances and giving equal treatment to joy and sorrow—is considered valuable in cultivating stable happiness.

Perseverance

Selfless service and hard work are valued as a means to attainment of goals.

On the whole, it appears that the indigenous constructs rooted in the Indian religio-philosophical tradition present a somewhat broader epistemological space to conceptualize human beings. Such concepts can provide a suitable alternative to look at the positive side of human life. Effort should be made to further explore these constructs so as to develop an indigenous model which can complement positive psychology perspective in vogue.

References

Ananda, P., and Prasad, A. (2011). Models of Personality in Buddhist Psychology. In R. M. M. Cornelissen, G. Misra, and Varma, S. (Eds.), *Foundations of Indian Psychology* (Vol. 1, pp. 146–69). New Delhi: Pearson.

Aquinas, T. (ca 1265 CE/1966). *Summa Theological* (Vol. 33, W. J. Hill, trans.). New York: McGraw-Hill.

Aspinwall, L. G., and Staudinger, U. M. (Eds.), (2003). *A Psychology of Human Strengths: Fundamental Questions and Future Directions for a Positive Psychology*. Washington, DC: American Psychological Association.

Averill, J. R., Catlin, G., and Chon, K. K. (1990). *Rules of Hope*. New York: Springer-Verlag.

Cornelissen, R. M. M., Misra, G., and Varma, S. (2011). *Foundations of Indian Psychology*. New Delhi: Pearson.

Csikszentmihalyi, M. (1990). *Flow: The Psychology of Optimal Experience*. New York: Harper & Row.

Csikszentmihalyi, M., and Nakamura, J. (2009). Psychological Capital. In S. J. Lopez (Ed.), *The Encyclopedia of Positive Psychology* (pp. 801–04). London: Wiley-Blackwell.

Cutuli, J. J., and Masten, A. S. (2009). Resilience. In S. J. Lopez (Ed.), *The Encyclopedia of Positive Psychology* (pp. 837–42). London: Wiley-Blackwell.

Dalal, A. S. (2001). *A Greater Psychology: An Introduction to the Psychological Thought of Sri Aurobindo*. Pondicherry: Sri Aurobindo Ashram.

Diener, E. (2000). Subjective Well-being: *The Science of Happiness and a Proposal for National Index*. American Psychologist, 55(1), 34–43.

Diener, E., Oishi, S., and Lucas, R. E. (2009). Subjective Well-being: *The Science of Happiness and Life Satisfaction*. In S. J. Lopez and C. R. Snyder (Eds.), *Oxford Handbook of Positive Psychology* (pp. 187–94). New York: Oxford University Press.

Duerlinger, J. (2008). Indian Buddhist Theories of Persons. In K. R. Rao, A. C. Paranjpe, and Ajit K. Dalal (Eds.), *Handbook of Indian Psychology* (pp. 129–41). New Delhi: Cambridge University Press.

Dwivedi, C. B. (2002). Ayurvedic Concepts of the Purusa, Dehaprakriti and Sattvaprakriti. In G. Misra and A. K. Mohanty (Eds.), *Perspectives on Indigenous Psychology* (pp. 305–25). New Delhi: Concept.

Edwards, L. M. (2009). Hope. In S. J. Lopez (Ed.), *The Encyclopedia of Positive Psychology* (pp. 487–91). London: Wiley-Blackwell.

Fredrickson, B. L. (2001). The Role of Positive Emotions in Positive Psychology: The Broaden-and-Build Theory of Positive Emotions. *American Psychologist*, *56*(3), 218–26.

Fredrickson, B. L. (2002). Positive Emotions. In C. R. Snyder and S. J. Lopez (Eds.), *The Handbook of Positive Psychology* (pp. 120–34). New York: Oxford University Press.

Fredrickson, B. L., and Joiner, T. (2002). Positive Emotions Trigger Upward Spirals Toward Emotional Well-being. *Psychological Science*, *13*(2), 172–75.

Fredrickson, B. L., Tugade, M. M., Waugh, C. E., and Larkin, G. R. (2003). What Good are Positive Emotions in Crisis? A Prospective Study of Resilience and Emotions Following the Terrorist Attack on the United States on September 11th, 2001. *Journal of Personality and Social Psychology*, *84*(2), 365–76.

Froh, J. J., Kashdan, T. B., Ozimkowski, K. M., and Miller, N. (2009). Who Benefits the Most From Gratitude Intervention in Children and Adolescents? Examining positive Effect as a Moderator. *Journal of Positive Psychology*, *4*(5), 408–22.

Handalsman, M. M., Knapp, S., and Gottlieb, M. (2002). Positive Ethics. In C. R. Snyder and S. J. Lopez (Eds.), *Handbook of Positive Psychology* (pp. 731–44). New York: Oxford University Press.

Isen, A. M. (1987). Positive Effect, Cognitive Processes, and Social Behaviour. In L. Berkowitz (Ed.), *Advances in Experimental Social Psychology*, (Vol. 20, pp. 203–53). New York: Academic Press.

Jain, J. P. (2008). Jaina Psychology. In K. R. Rao, A. C. Paranjpe and Ajit K. Dalal (Eds.), *Handbook of Indian Psychology* (pp. 55–72). New Delhi: Cambridge University Press.

Jeste, D. V., and Vahia, I. V. (2008). Comparison of the Conceptualization of Wisdom in Ancient Indian Literature with Modern Views: Focus on the Bhagavad Gita. *Psychiatry, 71*(3), 197–209.

Jung, C. G. (1933). *Modern Man in Search of a Soul*. London: Kegan Paul, Trench Trubner.

Kahneman, D. (1999) Objective Happiness. In D. Kanheman, E. Diener and N. Schwartz (Eds.), *Well-being: The Foundations of Hedonic Psychology* (pp. 3–25). New York: Russel Sage Foundation.

Keyes, C. L. M., and Haidt, J. (Eds.), (2003). *Flourishing: Positive Psychology and the Life Well Lived*. Washington, DC: American Psychological Association.

Langer, E. (2009a). Mindfulness versus Positive Evaluation. In S. J. Lopez and C. R. Snyder (Eds.), *Oxford Handbook of Positive Psychology* (pp. 279–94). New York: Oxford University Press.

Langer, E. (2009b). *Clockwise Mindful Health and the Power of Possibility*. New York: Random House.

Linley, P. A., and Joseph, S. (Eds.), (2004). *Positive Psychology in Practice*. Hoboken, NJ: Wiley Press.

Lopez, S. J. (Ed.) (2009). *The Encyclopedia of Positive Psychology*. London: Wiley-Blackwell.

Lopez, S. J., and Snyder, C. R. (Eds.), (2004). *Positive Psychological Assessment: A Handbook of Models and Measure*. Washington, DC: American Psychological Association.

Luthans, F. (2002a). The Need for and Meaning of Positive Organizational Behaviour. *Journal of Organizational Behaviour, 23*(6), pp. 695–706.

Luthans, F., Youssef, C. M., and Avolio, B. J. (2007). *Psychological Capital: Developing the Human Competitive Edge*. Oxford, UK: Oxford University Press.

Lyubomirsky, S., King, L., and Diener, E. (2005). The Benefits of frequent Positive Effect: Does Happiness Lead to Success? *Psychological Bulletin, 131*(6), pp. 803–55.

Massimini, F., and Delle Fave, A. (2000). Individual Development in a Bio-Cultural Perspective. *American Psychologist, 55*(1), 24–33.

Masten, A. S., and Reed, M. G. J. (2002). Resilience in Development. In C. R. Snyder and S. J. Lopez (Eds.), *Handbook of Positive Psychology* (pp. 74–88). New York, NY: Oxford University Press.

Masten, A. S., Cutuli, J. J., Herbers, J. E., and Reed, M. G. J. (2009). Resilience in Development. In S. J. Lopez and C. R. Snyder (Eds.), *Oxford Handbook of Positive Psychology* (pp. 117–31). New York: Oxford University Press.

Mayer, J. D., Lin, S. C., and Korogodsky, M. (2011). Exploring the Universality of Personality Judgments: Evidence From the Great Transformation (BCE 1000–BCE 200). *Review of General Psychology, 15*(1), 65–76.

Millon, T. (2012). On the history and Future Study of Personality and Its Disorders. *Annual Review of Clinical Psychology, 8*, 1–19.

Misra, G. (Ed.) (2011). *Handbook of Psychology in India*. New Delhi: Oxford University Press.

Murthy, P. K., and Salagame, K. K. K. (2007). The Concept of Triguna: A Critical Analysis and Synthesis. *Psychological Studies, 52*(2), 103–13.

Myers, D. G. (2000). The Funds, Friends and Faith of Happy People. *American Psychologist, 55*(1), 56–57.

Ong, A. D., and Van Dulmen, M. H. M. (2007). *Oxford Handbook of Methods in Positive Psychology*. New York: Oxford University Press.

Pande, N., and Naidu, R. K. (1992). Anasakti and Health: A Study of Non-Attachment. *Psychology and Developing Societies, 4*(1), 89–104.

Paranjpe, A. C., and Rao, K. R. (2008). Psychology in the Advita Vedanta. In K. R. Rao, A. C. Paranjpe, and A. K. Dalal (Eds.), *Handbook of Indian Psychology* (pp. 253–85). New Delhi: Cambridge University Press.

Park, N., Peterson, C., and Seligman, M. E. P. (2006). Character Strengths in Fifty-Four Nations and the Fifty US States. *Journal of Positive Psychology, 1*(3), 118–29.

Peterson, C. (2000). The Future of Optimism. *American Psychologist, 55*(1), 44–55.

Peterson, C., and Seligman, M. E. P. (2004). *Character Strengths and Virtues: A Handbook and Classification*. Washington DC: American Psychological Association.

Rao, K. R., Paranjpe, A. C., and Dalal, A. K. (2008). *Handbook of Indian Psychology* (pp. 253–85). New Delhi: Cambridge University Press.

Ryan, R. M., and Deci, E. L. (2000). Self Determination Theory and the Facilitation of Intrinsic Motivation, Social Development and Well Being. *American Psychologist, 55*(1), 68–78.

Salagame, K. K. K. (2011). Ego and Ahamkara: Self and Identity in Modern Psychology and Indian Thought. In R. M. M. Cornelissen, G. Misra and Varma, S. (Eds.), *Foundations of Indian Psychology* (pp. 133–145). New Delhi: Pearson.

Salagame, K. K. K., Raj, A., Murthy, K. P., Partimala, N., Rekha, K., and Gaur, S. (2005). Concept of Ahamkara: Theoretical and Empirical Analysis. In K. R. Rao (Ed.), *Towards a Spiritual Psychology: Essays in Indian Psychology*. New Delhi: Samvada Publishers.

Schwartz, B. (2000). Self Determination: The Tyranny of Freedom. *American Psychologist, 55*(1), pp. 79–88.

Seligman, M. E. P. (1991). *Learned Optimism*. New York: Knopf.

Seligman, M. E. P. (2002). *Authentic Happiness*. New York: Free Press.

Seligman, M., and Csikzentmihalyi, M. (2000). Positive Psychology. *American Psychologist*, *55*(1), pp. 5–13.

Sibia, A., and Srivastava, A. K. (2003). *Emotional Intelligence in the Indian Context: An Exploration.* Paper Presented at National Seminar on Psychology in the Indian Traditions: Conceptual and Methodological Issues for Indian Scholars. pp. 22–24. December, NCERT, New Delhi, India.

Singh, J. K. (2009). Emotional Intelligence: Theoretical and Cultural Perspectives. *Psychological Studies*, *54*(2), 157–59.

Singh, J. K., and Misra, G. (2000). Understanding Contentment in Everyday Life. *Indian Psychological Review*, *54* and *55* (4 Special Issue), 113–24.

Singh, J. K., and Misra, G. (2013). A Psycholexical Study of Personality Trait structure of Hindi-Speaking Indians. In Y. Kashima, E. Kashima, and R. Beatson (Eds.), *Steering the Cultural Dynamics: Selected Papers From the 2010 Congress of the International Association for Cross-Cultural Psychology* (pp. 154–59). Melbourne, Australi: International Association for Cross-Cultural Psychology.

Singh, J. K., Misra, G., and DeRaad, B. (2013). Personality Structure in the Trait Lexicon of Hindi, a Major Language Spoken in India. *European Journal of Personality*, *27*(6), 605–20.

Snyder, C. R. (2000a). Genesis: The Birth and Growth of Hope. In C. R. Snyder (Ed.), *Handbook of Hope: Theory, Measures, and Applications* (pp. 25–38). San Diego, CA: Academic Press.

Snyder, C. R. (2000b). *Handbook of Hope: Theory, Measures, and Applications*. San Diego, CA: Academic Press.

Snyder, C. R., and Lopez, S. J. (Eds.), (2002). *Handbook of Positive Psychology*. New York: Oxford University Press.

Snyder, C. R., Irving, L., and Anderson, J. (1991). Hope and Health: Measuring the Will and the Way. In C. R. Snyder and D. R. Forsyth (Eds.), *Handbook of Social and Clinical Psychology* (pp. 285–305). Elmsford, NY: Pergamon Press.

Srivastava, A. K., and Misra, G. (2001). Lay People's Understanding and Use of Intelligence: An Indian Perspective. *Psychology and Developing Societie, 13*(1), 25–49.

Srivastava, A. K., and Misra, G., (2003). Going Beyond the Model of Economic Man: An Indigenous Perspective on Happiness. *Journal of Indian Psychology, 21*(1), pp. 12–29.

Stajkovic, A. D., and Luthans, F. (1998). Social Cognitive Theory and Self-efficacy: Going Beyond Traditional Motivational and Behavioural Approaches. *Organizational Dynamics*, *26*(4), 62–74.

Stocks, E. L., and Lishner, D. A. (2009). Empathy. In S. J. Lopez (Ed.), *The Encyclopedia of Positive Psychology* (pp. 320–26). London: Wiley-Blackwell.

Strand, E. B., Reich, J. W., and Zautra, A. J. (2009). Positive Experience. In S. J. Lopez (Ed.), *The Encyclopedia of Positive Psychology* (pp. 721–26). London: Wiley-Blackwell.

Taylor, S. E., Kemeny, M. E., Reed, G. M., Bower, J. E., and Gruenwald, T. L. (2000). Psychological Resources, Positive Iillusion and Health. *American Psychologist*, *55*(1), 99–109.

Terman, L. M. (1939). The Gifted Student and His Academic Environment. *School and Society.* 49, 65–73.

Terman, L. M., Buttenwieser, P., Ferguson, L. W., Johnson, B. W. and Wilson, D. P. (1938). *Psychological Factors in Marital Happiness*. New York: McGraw-Hill.

Vaillant, G. E. (2000). Adaptive Mental Mechanisms: Their Role in Positive Psychology. *American Psychologist*, *55*(1), 89–98.

Varma, S. (2011). Integral Psychology: A New Science of Self and Personality. In R. M. M. Cornelissen, G. Misra, and Varma, S. (Eds.), *Foundations of Indian Psychology* (pp. 170–97). New Delhi,Pearson.

Watson, J. (1928). *Psychological Care of Infant and Child.* New York: Norton.

Zautra, A. J., Potter, P. T., and Reich, J. W. (1997). The Independence of Affect is Context Dependent: An Integrative Model of the Relationship Positive and Negative Effect. In K. W. Schaie and M. P. Lawton (Eds.), *Annual Review of Gerontology and Geriatrics* (Vol. 17, pp. 75–103). New York: Springer.

CHAPTER 3

Virtues and Well-being

Surendra Kumar Sia

Department of Applied Psychology, Pondicherry University, Pondicherry

Abstract Virtue is a positive trait or quality considered to be morally respectable and, thus, is honoured as the foundation for good moral being. Virtue is an attribute of a person which leads to moral excellence and collective well-being. Virtues have been described in a variety of religious, philosophic and cultural traditions. The present chapter would begin with a historical perspective of virtues, having focus upon Eastern traditions. Confucianism advocates five virtues which are central to lead a moral life. Buddhism, a major Eastern philosophy, gives an important place to virtues and describes those as *Brahma Vihars*. Stalker (1902) mentions about seven cardinal virtues, which in fact is an union of two sets of catholic virtues comprising four cardinal virtues and three theological virtues. The chapter then attempts to discuss the scientific base of virtue, in terms of strengths, as grounded by Peterson and Seligman (2004). Though one can certainly be virtuous without being religious, still religion can provide a base for studies on morality and virtue. Therefore, the next section of the chapter deals with the relationship between religion and virtue. The chapter also explicates the role of virtues towards psychological well-being. The chapter ends with future research implications for studies on virtue as a construct.

Virtues and Well-being

Virtue is a positive trait or quality considered to be morally respectable and, thus, is honoured as the foundation for a good morally upright human being. A virtue is an attribute of a person which leads to moral excellence and collective well-being. Virtue is the capacity to perform what is good. With virtue, our mind becomes our own. A major emphasis of positive psychology is to identify as well as foster the bases of a well-lived life. The characteristics of a good life are intimately associated with virtue. Virtue is positive in its own regard. We usually admire people who are quite sociable, outgoing, optimistic, etc. We also admire people who exhibit virtuous qualities like kindness, honesty, gentleness, reverence, integrity and courage. These qualities not only have significance for individuals, but also have greater value for the society. Usually, these qualities have moral as well as spiritual connotations. Virtue is a meeting point of divine perfection and human life as an ideal state. Confronting the ideal prototype, man finds himself face-to-face not with a moral 'must' but with an 'is'. Virtue stands, thus 'midway' between God and moral imperatives (Lindbom, 1975). Despite being a positive asset for the individual as well as the society, for a long time, virtue has not been considered as an appropriate construct for systematic psychological query. The study of virtue was assumed to be easily biased by researcher's moral beliefs and prevailing cultural norm (Tjeltveit, 2003). However, a renewed interest has started to emerge since the positive psychologists realize that without the inclusion of this moral dimension, explanation of human behaviour may remain incomplete. Before travelling on the scientific expositions related to virtues, let us project briefly an historical account from the Eastern perspective.

Virtues and Buddhism

The Eastern philosophies and religious practices including Buddhism and Confucianism provide the insight regarding how to lead a harmonious, simplistic and fulfilling lifestyle. All these insights can be applied in everyday life irrespective of one's religious beliefs. The practices prescribed in Buddhism are a progressive list of virtues. The four *brahmavihāras*, which are considered to be above the universal virtues, occupy an important place in Buddhist tradition (Wetlesen, 2002; Bodhi, 2000). Cultivation of the four virtues, which are also called, 'four immesurables (*apramāna*)', makes the human being develop wholesome attitudes towards others in society and to contribute towards greater happiness. These virtues are: *maitri* (loving-kindness or benevolence), *karunā* (compassion), *mudita* (empathic joy) and *upeksha* (equanimity) (Sangharakshita, 1991).

Maitri or loving-kindness counters ill will. The attitude of loving-kindness is like the feeling which a mother has for her newborn. However, this should not remain confined to those for whom we have some attachment. It should also extend to others whom we may know faintly or not know at all. In a nutshell, it is the wish that all conscious beings, without any exception, be happy. *Karunā* or compassion counters cruelty. It makes the heart move at the pain of others. The individual wishes to crush and destroy the pain of others and, also helps and embraces the distressed people. We can inculcate the attitude of compassion in the world around us. *Mudita* or empathic joy is the attitude of rejoicing in the happiness and virtues of others. It counters jealousy and makes people less self centered. We can experience this appreciative joy even in our daily lives. It is especially the vicarious joy which comes from delighting in other people's well-being. It should not be confounded with pride. The individual experiencing *mudita* must not have anything to gain from other's accomplishments. If we are happy when others are healthy and prosperous, it is called *mudita*. *Upeksha* or equanimity is the virtue which can lead to the Buddhist path of *nirvana*. It refers to mental and emotional stability in the face of the worldly fluctuations.

Equanimity is a protection from the 'eight worldly winds': praise and blame, success and failure, pleasure and pain, fame and disrepute (Fronsdal, 2004). Moreover, it counters clinging as well as aversion. In other words, we should try to remain free from all points of self-reference and regard all living beings as equals, irrespective of their present relationship with us. However, it should not be confused with indifference, which implies the lack of concern for others. True equanimity is the culmination point of the four virtues advocated by the Buddhist tradition. Buddhism also advocates the *pāramitās*, which refer to the perfection or culmination of certain virtues and can be cultivated as a way of purification (Bodhi, 2005). These are: *dana* (generosity, i.e., giving of oneself), *sila* (virtue, morality, proper conduct), *nekkhamma* (renunciation), *panna* (transcendental wisdom, insight), *viriya* (energy, diligence, vigour), *khanti* (patience, tolerance, acceptance, endurance), *sacca* (truthfulness, honesty) and *adhitthana* (determination, resolution). All these virtues are also respected by Buddhists as powerful counteragents against negative mental states like anger, greed and pride.

Confucianism and Virtues

Confucianism, like other eastern religions, provides basic insight regarding how to lead a simple, harmonious and fulfilling lifestyle. The major focus is upon humans and the fundamental principles of humanity. Confucian moral thought is a role-based virtue ethics. It is role-based because 'it is based on the roles that make an agent the person he or she is' (Nuyen, 2009). According to Confucianism, people must continue a life-long journey of learning and self-discovery to become the ideal person. Confucianism does not glorify allegiance to divine wish or higher law (Juergensmeyer, 2005). The belief is that human beings can be taught and edified through personal and collective effort. Confucian tradition emphasizes upon cultivation of virtue and maintenance of ethics. Individuals should not only learn certain virtues, they must practice and make such virtues part of their character or pattern of life. The attainment of virtue is at the core of Confucian teachings (Snyder and Lopez, 2007). The important virtues advocated are: *ren, yi* and *li.*

Ren (*Jen*) is the core virtue as per the teachings of Confucius. *Ren* (Humaneness) refers to altruism and humaneness towards other individuals. This is well comprehended in the Confucian version of the ethics of reciprocity. One should treat others as one would like others to treat oneself (Flew, 1979). *Ren* has both a broad and a narrow perspective (Shun, 1997). In the narrow sense, it can be termed as *benevolence*, which means, caring for or loving others. This is associated with the natural affective component of the individual. The broad sense of *ren* can be translated as perfect virtue or humanness. This broad sense is the unity of all the virtues including *ren* in the narrow sense. A similar type of classification is put forth by Luo (2011) addressing different understandings of *ren. Ren* as a first-order particular virtue is one among many virtues. Further, *ren* as the higher-order general virtue is an umbrella term used to cover a variety of virtues. It is used as an umbrella term to cover five virtues: respect (*gong*), tolerance (*kuan*), trustworthiness (*xin*), diligence (*min*) and kindness (*hui*) (Luo, 2011).

The virtue of *li,* which refers to propriety or etiquette, gives clear instruction for proper behaviours on the part of the individuals based on their respective roles in the society. *Li* was constituted by the rules specifying rituals governing sacrifices performed by individuals or groups in honour of their common ancestors (Chang, 1983). Afterwards, the scope of *li* has been stretched to include etiquette and appropriate manners in social interactions. *Li* can provide a proper common ground for social interactions and overcome the problem of chaos produced by unregulated natural desires. Moreover, by regulating resource distribution, *li* can produce prosperity (Xunzi, 2001). According to the Confucian perspective, *li* has an important role in the practice of some virtues. It places standards to exhibit proper manner and experience right emotion in specific context of social interaction. Without individual's observance of *li*, some virtues can become vices.

Yi, translated as *righteousness, appropriateness* or *dutifulness,* functions as an intermediary between *ren* and *li*. This virtue is a character trait related to the motivation to abide by *li* or social norms (Cua, 2007).*Yi* is connected with the emotions of shame and aversion arising out of the negative emotion triggered by failure to conform to *li*. (van Norden, 2007). Some behaviours ought to be exhibited due to the reason that those are right irrespective of their outcomes. The core of the act is its rightness regardless of its intention or consequence. Therefore, *yi* is a different way than *stoicism* (intention with determinism) or *utilitarianism* (maximizing consequences). Confucians also vouch for other virtues like *wen* (appreciation of various types of art), *xiao* (Filial Piety), *shu* (being mindful about how one's actions will affect other people) and *zhi* (knowledge) (Molloy, 2010).

Virtues: Hindu Perspectives

Unlike other traditions, Hinduism does not seem to have a specific founder. The Hindu theory of life runs through three great ages. The first is the Vedic Age—the age of affirmation. The second is the age of the Upanishads—the age of denial of the world and affirmation of the spirit. The third is the age of synthesis when the values of the world are reaffirmed in the light of the spirit. This was the age of the Gita. It was in this way that the Hindu theory gains such a depth and complexity. The world is affirmed, denied and re-affirmed. The Hindu way of life or *Sanatana Dharma*, as it was originally called, is inclusive by nature (Buch, 2012). Sankara's *Vivekacudamani* (verses 18–28) suggests the human being to prepare for self-realization through acquisition of four means (*sadhna-chatustaya*) (Paranjpe and Rao, 2008). Out of these four means, attainment of *shad-sampat* or six virtues is most important. The virtues are control of the mind, tranquility and calmness (*shama*), control of the sense organs (*dama*), inner withdrawal or renunciation of worldly activities (*uparati*), patient forbearance of suffering and endurance of changing situations (*titiksha*), unconditional faith and trust faith in the teacher, the *atman* and the scriptures (*shraddha*) and concentration of the mind, study and contemplation (*samadhana*) (Puligandla, 1985).

The Bhagavad Gita, a major Hindu text, depicts an integral philosophy of life and transformation. This holy book enlists twenty-six divine qualities of which the edifice of righteous action is built (Sinha, 2005). These virtues help the individual to reach the supreme destination. Perhaps, all the important virtues which have become significant constructs of research by present-day positive psychologists are covered in the first three stanzas of the sixteenth chapter of the Bhagavad Gita.

Abhayam sattva-samsuddhir jnana-yoga-vyavasthitih
Danam damas ca yajnas ca svadhyayas tapa arjavam
(Bhagavad Gita, 16.1)

Ahimsa satyam akrodhas tyagah santir apaisunam
Dayabhutesv aloluptvam mardavam hrir acapalam
(Bhagavad Gita, 16.2)

Tejah ksama dhrtih saucam adroho natimanita
Bhavanti sampadam daivim abhijatasya bharata
(Bhagavad Gita, 16.3)

We may describe some of the important ones. *Abhayam* is fearlessness due to the absence of anxiety which arises from the dread of harm to the physical body or the prospect of losing what is precious. *Danam* is charity offered to worthy recipients. *Damah* refers to self-restraint and controlling the mind without being influenced by sensory stimuli. *Ahimsa* is non-violence to, or not hurting any living being by thought, word or action. This virtue has been adopted, practiced and advocated successfully by Mahatma Gandhi. *Satyam* is truthfulness or showing the courage to stand up and voice the truth for everybody's benefit. *Akrodha* means freedom from anger and aggression based upon lack of resentment for others. *Tyagah* is renunciation or abandonment of things opposed to self-realization. *Santih* implies peace and tranquility through maintaining calm in sense organs. *Daya* refers to compassion, sympathy and empathy for the misery of others. It is one of the vital virtues in spiritual development. *Arjava* refers to being honest, both with others as well as ourselves. We should renounce deception. *Hrih* means being modest and feeling uneasy for own inappropriate behaviour. We should foster enough modesty so that we can apologize for our mistakes in the presence of others. *Kshama* is forgiveness or compassion. It is important to be tolerant and patient towards others and attempt to understand them. *Dhritih* is fortitude and ability to persevere in spite of hardships and adversary conditions. We should constantly follow the righteous path even in the face of great difficulties (Bhagavad Gita 16.1, 2 and 3).

Virtues: Empirical Orientation

Deliberations upon positive traits would be incomplete if we do not include positive human qualities which are admired morally and ethically. We respect individuals who exhibit strengths of character which manifests virtuous attributes like honesty, integrity and kindness, etc. Virtue is esteemed positively by everybody because of its value in society and its strong association with the religious and secular mores of life. However, in spite of its relevance to the individual as well as the society, virtue has not been considered as an appropriate construct for scientific investigation in the history of psychology (Baumgardner and Crothers, 2009). The study of virtue was assumed to be easily tainted and biased by the moral beliefs of researchers and existing cultural practices (Tjeltveit, 2003). Virtue and character were some of the dominant topics in the early periods of psychology, and character development was considered as an important aspect (Leahey, 1994; Sussman, 1973). But these concepts fell into disinterest as psychologists attempted to segregate scientific fact from moral value in their attempt to become an accepted scientific branch (Nicholson, 1998; Ross and Nisbett, 1991). Personality psychologists like Allport have consistently attempted to exclude the evaluative components from the discussion of personality and devalued the importance of character and virtue to become a part of psychology. They considered character and virtue to be normative components of ethics unlike the scientific personality traits (Allport, 1921; Allport and Vernon, 1930). However, it is not easy to separate moral attributes from the description of personality. Virtue and character are integral to the widely acclaimed five factor model of personality. Two of those five factors, viz., agreeableness and conscientiousness, are clearly related to virtue (Costa and McCrae, 1991; McCrae and John, 1992).

The other influential factor responsible for the disfavour shown by psychologists towards virtue is the publication of Hartshorne and May's studies on honesty among schoolchildren (Hartshorne and May, 1928; Hartshorne, May and Maller, 1929; Hartshorne, May and Shuttleworth, 1930).They found that children were dishonest if got the opportunity. Their conclusion was that situational contributors were more important than character for manifestation of moral behaviour. However, these findings have been reinterpreted later theoretically and empirically (Fowers, 2012). Virtue theory does not indicate that majority of the individuals, including children, will act virtuously across the situations. The finding that some children acted honestly in spite of incentives for not being honest, suggests the presence of the strength of character. Statistical re-analyses also cast doubt on the conclusions by Hartshorne and May (Burton, 1963; Epstein and O'Brien, 1985; Rushton, 1984). Unlike the previous dismissals by psychologists, Baumeister and Exline (1999) argue that virtue concepts can be subjected to empirical research with available methods. The advent of positive psychology, with its interest in virtue has also contributed to this literature (Peterson and Seligman, 2004). More and more empirical literature is emerging on global virtue (Cawley, Martin and Johnson, 2000; Hawkins, Fowers, Carroll and Yang, 2006; Steger, Hicks, Kashdan, Krueger and Bouchard, 2007; Shryack, Steger, Krueger and Kallie, 2010; Walker and Pitts, 1998) and specific virtues like forgiveness (McCullough, Pargament and Thoresen, 2000; Mullet, Girard and Bakhshi, 2004) gratitude (Emmons and Crumpler, 2000; McCullough, Emmons and Tsang, 2002), courage (Pury and Kowalski, 2007) and practical wisdom (Kramer, 2000; Staudinger, Lopez and Baltes, 1997).

Psychologists have attempted to delineate definitions for virtue. Broadie (1991) defines that excellence or virtue is nothing but a characteristic which makes the difference between functioning and functioning well. It is merely a psychological process that enables a person to think and act so as to benefit himself or herself and society (McCullough and Snyder, 2000). Virtues are broad-band, socially desirable, individual difference constructs that are valued across cultures (Peterson and Seligman, 2004). Virtues are core characteristics valued by moral philosophers and character strengths are less abstract personality traits which may be used to achieve virtues (Carr, 2004). Virtues are simply human excellences or character strengths that make it possible for individuals to pursue characteristically human good that allow them to flourish as human beings (Fowers, 2012). Beyond definition, psychologists have put sincere effort in identifying the nature and structure of virtue in different contexts and different populations (Cawley, Martin and Johnson, 2000; Dahlsgaard, 2005; Park and Peterson, 2006; Peterson and Park, 2004; Peterson, Park, Pole, D'Andrea and Seligman, 2008; Van Eeden, Wissing, Dreyer, Park and Peterson, 2008; Shryack, Steger, Krueger and Kallie, 2010). Except the factor analysis carried out by Cawley, *et al.* (2000), all other researchers have basically undertaken their exploration using the twenty-four character strengths enumerated in virtues in action (VIA) model. This VIA model is originally based upon comprehensive review of literature and professional consensus and distinctions have been made between virtue, strengths and enabling themes (Dahlsgaard, Peterson and Seligman, 2005; Peterson and Seligman, 2004). Character strengths are routes for achieving virtues (Peterson and Seligman, 2004). The twenty-four strengths are organized around six virtues. Each virtue can be defined in terms of a number of character strengths which represent the ingredients for developing virtue. The six virtues in this model are wisdom, courage, humanity, justice, temperance and transcendence.

Wisdom refers to acquisition and use of knowledge. This can be achieved through the strengths of creativity, curiosity, taking wider perspective of the world, critical thinking and interest to learn new things. It is not necessarily related with formal education or intelligence quotient. Intelligence provides the basic knowledge for daily life tasks, whereas wisdom consists of the know-how, judgement and flexibility to solve predominant life problems for the common good (Clayton, 1982; Sternberg, 1985). Courage is the emotional strength and will power to accomplish goals in spite of internal as well as external adversity and opposition. This virtue can be fostered through perseverance, bravery, honesty and the feeling of being alive and excited. Similarly, researchers identified three types of courage, viz., physical courage, moral courage and vital courage in similar model (O'Byrne, Lopez and Peterson, 2000). Physical courage includes the maintenance of societal good through the manifestation of physical behaviour. Moral courage refers to expression of authenticity in the face of disapproval or rejection. Vital courage means perseverance through a disease or disability even when the outcome is ambiguous (Snyder and Lopez, 2007). Humanity is the virtue of sympathy, empathy, compassion and love in interpersonal relationship. To develop this virtue, we need to understand the social world, be generous and kind towards others and need to value close relationships. Justice is a vital component for healthy community life. It involves the ability to relate effectively within the society and can be achieved by fairness, teamwork and organizing group activity. The virtue of temperance is a kind of self-awareness which protects against expression of excess. This can be inculcated through self-regulation, modesty and by choosing actions with care. Temperance also involves the ability to forgive the hurtful behaviours of others. The virtue of transcendence raises the individual above the preoccupations of daily life and provides a deeper meaning of life through a connection with the wider universe. This can be developed through gratitude, hope, spirituality and appreciation of beauty and excellence. Probably all six of these virtues must be present at the above threshold level for a person to be considered as of good character (Carr, 2004). Although, a number of researches have been carried out using the character strengths embedded in VIA model, there is not enough evidence to substantiate the six virtues structure (Park and Peterson, 2005; Dahlsgaard, 2005; Park and Peterson, 2008; Peterson, Park, Pole, D'Andrea and Seligman, 2008; Haidt, 2008; Shryack, Steger, Krueger and Kallie, 2010). An important objective of the positive psychology movement has become to describe and evaluate virtuous personality for the benefit of individuals and society (Peterson and Park, 2004; Peterson and Seligman, 2004). More researches may be carried out for establishing empirically derived dimensions of virtue to fulfil this objective.

In addition to the attempt for delineating global virtues like the ones mentioned here, researchers are also focusing upon specific virtues like forgiveness (McCullough, Pargament and Thoresen, 2000; Mullet, Girard and Bakhshi, 2004) gratitude (Emmons and Crumpler, 2000; McCullough, Emmons and Tsang, 2002), courage (Pury and Kowalski, 2007), and practical wisdom (Kramer, 2000; Staudinger, Lopez and Baltes, 1997).

Virtue and Well-being

Positive psychologists should try to explicate the relationship of the concept or construct they focus with well-being. Virtue, being a major topic of discussion in religious scriptures and philosophical texts, has definitely a strong association with well-being or good life. Therefore, MacIntyre (1984) mentions that the concept of *the good life* is prior to the concept of a virtue. Virtues make it possible to pursue the good life. But, unfortunately, the psychological literatures on the topics of virtue and good life are extremely fragmented, lacking cohesion in either domain and exhibiting precious few connections between these two growing and deeply interconnected areas of interest (Fowers, 2010, 2012). Psychologists, who take up studies on virtue, should link it to the concept of well-being. When we say, 'well-being', it is not simply the hedonic happiness, i.e., pleasantness. Being in positive psychology, we definitely look forward to the human being to be happy. But, at the same time, we should also emphasize whether the individual is realizing his/her potential to lead an authentic life or not. Aristotle defines a term called 'eudaimonia' as realizing one's full potential. Therefore, we need to view well-being with multiple foci. In this context the concept of complete mental health (Ryff and Keyes, 1995) is worth mentioning. They view complete mental health as combination of emotional well-being, social well-being and psychological well-being. Emotional well-being, referred as subjective well-being, can be defined as the presence of positive affect, absence of negative affect and overall life satisfaction. The components of psychological well-being are self-acceptance, personal growth and purpose in life, environmental mastery, autonomy and positive relations. Social well-being involves social coherence, feeling of being part of community, social contribution and social acceptance. Complete mental health consists of high levels of emotional, social and psychological well-being (Keyes and Lopez, 2002). Attempts can be made to investigate the contribution of virtue towards these different facets of well-being in different spheres of life. A number of practical applications of virtue have been advocated in varied areas like education (Park, Peterson and Seligman, 2006; Steen, Kachorek and Peterson, 2003), clinical settings (Seligman and Peterson, 2003; Seligman, Steen, Park and Peterson, 2005) and organizations (Peterson and Park, 2006).

Human virtues make sense within the relational context of social functions and purposes. The appropriation of virtue ethics in psychotherapy can be seen as a significant force in the state-of-art progress that is taking place among a growing number of psychiatrists and psychologists who are gradually becoming critical of the ideology of instrumentalism and technicism in their own clinical practice (Aho, 2012). Modern culture and modern science have been hampered by materialism, dualism, realism and idealism. Though all things are thought to be real and fundamental, they are in fact abstracted from their contexts (Slife and Richardson, 2008). Slife and associates (e.g., Slife and Richardson, 2008) established a therapeutic school exclusively on strong relationality, which led them to employ Aristotle's virtue ethics. They mention that virtues can only be acquired as character strengths through the lived activities. The participant's aspirations towards virtues lead to eudaimonic well-being (Slife, 2012). An extensive study by Park, Peterson and Seligman (2004), reveals that consistently and robustly associated with life satisfaction are hope, zest, gratitude, love and curiosity. Moreover, excess of any one character strength does not diminish life satisfaction. Employees scoring high on virtues are more satisfied and engaged at work, indicate higher levels of psychological well-being and perceive higher levels of functioning and performance (Nelson and Cooper, 2007; Bakker and Schaufelli, 2008; Burke, Ng and Fiksenbaum, 2009). We can also discuss about the contribution of specific virtues towards different facets of well-being. Reviews suggest that forgiveness leads to small but consistent positive outcomes in health and well-being (McCullough and Witvliet, 2002; Worthington, 2006), acts as antidote against the negative effects of hostility (Witvliet, Ludwig and Vander Laan, 2001), improves interpersonal relationships (Karremans and Van Lange, 2004; McCullough, Rachal, Sandage, Worthington, Brown and Hight, 1998; Tsang, McCullough, and Fincham, 2006) and adjustment of women with breast cancer (Glinder and Compas, 1999; Romero, Friedman, Kalidas, Elledge, Chang, and Liscum, 2006). Another widely researched virtue, which is found to have significant association with well-being, is gratitude. Studies indicate that gratitude make us feel happy, contended and joyful (Bono, Emmons and McCullough, 2004; Emmons and McCullough, 2003), lowers risk for major depression, generalized anxiety disorder and phobia (Kendler, Liu, Gardner, McCullough and Larsen, 2003), benefits people with post traumatic stress disorder(Kashdan, Uswatte, and Julian, 2006), increases post-traumatic growth (Peterson and Seligman, 2003), promotes relationship formation and maintenance (Algoe, Haidt, and Gable, 2008) and is integral to positive functioning (Maltby, Day and Barber, 2004).

Virtue: Research Implications

Virtue is an emerging construct in the field of positive psychology. However, it has a long moral and religious background. In all the religions adapted in different cultures, a major focus is given on virtuous behaviour. Psychologists from different cultural backgrounds may attempt to operationalize as well as empirically examine the different terms and concepts suggested in various religions related to virtue. In the present context, an important objective of the positive psychology movement has been to elaborate or evaluate virtuous behaviour for the benefit of the individuals and society (Peterson and Park, 2004; Peterson and Seligman, 2004). Therefore, adequate research is required to establish empirically derived dimensions of virtue. Future research is also needed to identify the mechanisms through which virtue relates to subjective well-being and psychological well-being. It is the positive sign of a developing field, when researchers orient their attention from establishing the effect to analysing why that effect takes place. Positive psychology research seems to be in that position. In most of the researches covered, virtue relates to variables which seem to have broad impacts on individual's life. Additional researches are required to investigate whether virtue is related to subsidiary outcomes through these mechanisms. Similarly, further research is needed to delineate precise cognitions as well as cognitive mechanisms which may explain the way character strengths of virtues operate. The development of virtue and character strength in different forms is also another interesting area of research. Moreover, the available literature does not show any research which has examined whether any negative side is also associated with virtue. Yet there could be. Therefore, researchers may also undertake studies to identify the conditions under which virtue may become maladaptive.

References

Aho, K. A. (2012). Assessing the Role of Virtue Ethics in Psychology: A Commentary on the Work of Blaine Fowers, Frank Richardson, and Brent Slife. *Journal of Theoretical and Philosophical Psychology, 32*(1), 43–49.

Allport, G. W. (1921). Personality and Character. *Psychological Bulletin, 18*(9), 441–55.

Allport, G. W., and Vernon, P. E. (1930). The Field of Personality. *Psychological Bulletin, 27*(10), 677–730.

Algoe, S. B., Haidt, J., and Gable, S. L. (2008). Beyond Reciprocity: Gratitudeand Relationships in Everyday Life. *Emotion, 8*(3), 425–29.

Bakker, A. B., and Schaufeli, W. B. (2008). Positive Organizational Behavior: Engaged Employees in Flourishing Organizations. *Journal of Organizational Behavior, 29*(2), 147–54.

Baumgardner, S. R., and Crothers, M. K. (2009). *Positive Psychology.* New Delhi: Pearson Education.

Baumeister, R. F., and Exline, J. J. (1999). Virtue, Personality, and Social Relations: Self-control as the Moral Muscle. *Journal of Personality, 67*(6), 1165–94.

Bodhi, B. (2000). *Facing the Future.* Kandy, Srilanka: Buddhist Publication Society.

Bodhi, B. (2005). A Treatise on the Paramis from the Commentary to the Cariyapitaka by Acariya Dhammapala Translated from the Pali. Retrieved from: http://www.accesstoinsight.org /lib.

Bono, G., Emmons, R. A., and McCullough, M. E. (2004). Gratitude in Practice and the Practice of Gratitude. In P. A. Linley, and S. Joseph (Eds.), *Positivepsychology in Practice.* Hoboken, NJ: John Wiley and Sons, Inc.

Broadie, S. (1991). *Ethics with Aristotle.* Oxford, UK: Oxford University Press.

Buch, M. A. (2012). *The Principles of Hindu Ethics.* Baroda: Srimatham.com

Burke, R. J., Ng, E. S. W., and Fiksenbaum, L. (2009). Virtues, Work Satisfactions and Well-being Among Nurses. *International Journal of Workplace Health Management, 2*(3), 202–19.

Burton, R. V. (1963). Generality of Honesty Reconsidered. *Psychological Review, 70*(6), 481–99.

Carr, A. (2004). *Positive Psychology. The Science of Happiness and Human Strengths*. London: Brunner-Routledge

Cawley, M. J., III, Martin, J. E., and Johnson, J. A. (2000). A Virtues Approach to Personality. *Personality and Individual Differences, 28*(5), 997–1013.

Chang, K. C. (1983). *Art, Myth, and Ritual: The Path to Political Authority in Ancient China*. Cambridge, MA: Harvard University Press.

Clayton, V. (1982). Wisdom and Intelligence: The Nature and Function of Knowledge in Later Years. *International Journal of Aging and Human Development, 15*(4), 315–21.

Costa, P. T., Jr., and McCrae, R. R. (1991). Facet Scales for Agreeableness and Conscientiousness: A Revision of the NEO Personality Inventory. *Personality and Individual Differences, 12*(9), 887–98.

Cua, A. S. (2007). Virtues of Junzi. *Journal of Chinese Philosophy, 34* (Suppl 1), 125–42.

Dahlsgaard, K. (2005). Is Virtue More than Its Own Reward? *Dissertation Abstracts International: Section B: The Sciences and Engineering, 66*(6-B) (UMINo.AAI3179723).

Dahlsgaard, K., Peterson, C., and Seligman, M. E. P. (2005). Sharedvirtue: The Convergence of Valued Human Strengths Across Culture and History. *Review of General Psychology, 9*(3), 203–13.

Emmons, R. A., and Crumpler, C. A. (2000). Gratitude as a Human Strength: Appraising the Evidence. *Journal of Social and Clinical Psychology, 19*(1), 56–69.

Emmons, R. A., and McCullough, M. E. (2003). Counting Blessings versus Burdens: Experimental Studies of Gratitude and Subjective Well-being. *Journal of Personality and Social Psychology, 84*(2), 377–89.

Epstein, S., and O'Brien, E. J. (1985). The Person-Situation Debate in Historical and Current Perspective. *Psychological Bulletin, 98*(3), 513–37.

Flew, A. (1979). *Golden rule: A Dictionary of Philosophy*. London: Pan Books.

Fowers, B. J. (2010). *Aristotle on Eudaimonia: On the Virtue of Returning to the Source.* Unpublished manuscript.

Fowers, B. J. (2012). Placing Virtue and the Human Good in Psychology. *Journal of Theoretical and Philosophical Psychology, 32*(1), 1–9.

Fronsdal, G. (2004). *Equanimity*. Retrieved from www.insightmeditationcenter.org.

Glinder, J. G., and Compas, B. E. (1999). Self-blame Attributions in Women with Newly Diagnosed Breast Cancer: A Prospective Study of Psychological Adjustment. *Health Psychology, 18*(5), 475–81.

Haidt, J. (2008). Morality. *Perspectives on Psychological Science, 3*(1), 65–72.

Hartshorne, H., and May, M. A. (1928). *Studies in the Nature of Character (Vol. 1): Studies in Deceit.* New York, NY: Macmillan.

Hartshorne, H., May, M. A., and Maller, J. B. (1929). *Studies in the Nature of Character (Vol. 2): Studies in Service and Self-Control.* New York, NY: Macmillan.

Hartshorne, H., May, M. A., and Shuttleworth, F. K. (1930). *Studies in the Nature of Character (Vol. 3): Studies in the Organization of Character.* New York, NY: Macmillan.

Hawkins, A. J., Fowers, B. J., Carroll, J. S., and Yang, C. (2006). Conceptualizing and Measuring Marital Virtues. In S. Hofferth and L. Casper (Eds.), *Handbook of Measurement Issues in Family Research* (pp. 67–83). Hillsdale, NJ: Erlbaum.

Juergensmeyer, M. (2005). *Religion in Global Civil Society*. Oxford: Oxford University Press.

Karremans, J., and Van Lange, P. A. M. (2004). Back to Caring after Being Hurt: The Role of Forgiveness. *European Journal of Social Psychology, 34*(2), 207–27.

Kashdan, T. B., Uswatte, G., and Julian, T. (2006). Gratitude and Hedonic and Eudaimonic Well-being in Vietnam War Veterans. *Behavior Research and Therapy, 44*(2), 177–99.

Kendler, K. S., Liu, X. Q., Gardner, C. O., McCullough, M. E., Larson, D., and Prescott, C. A. (2003). Dimensions of Religiosity and their Relationship to Lifetime Psychiatric and Substance Use Disorders. *American Journal of Psychiatry, 160*(3), 496–503.

Keyes, C. L. M., and Lopez, S. J. (2002). Toward a Science of Mental Health: Positive Directions in Diagnosis and Treatment. In C. R. Snyder and S. Lopez (Eds.), *The Handbook of Positive Psychology* (pp. 45–59). New York: Oxford University Press.

Kramer, D. A. (2000). Wisdom as a Classical Source of Human Strength: Conceptualization and Empirical Inquiry. *Journal of Social and Clinical Psychology, 19*(1), 83–101.

Leahey, T. H. (1994). *A History of Modern Psychology, 2nd Ed.* Englewood Cliffs, NJ: Prentice Hall.

Lindbom, T. (1975). Virtue and Morality. *Studies in Comparative Religion, 9*(4), 227–35.

Luo, S. (2011). Is *Yi* More Basic than *Ren* in the Teachings of Confucious? *Journal of Chinese Philosophy, 38*(3), 427–43.

MacIntyre, A. (1984). *After Virtue: A Study in Moral Theory.* Notre Dame, IN: University of Notre Dame Press.

Maltby, J., Day, L., and Barber, L. (2004). Forgiveness and Mental Health Variables: Interpreting the Relationship Using an Adaptational–Continuum Model of Personality and Coping. *Personality and Individual Differences, 37*(8), 1629–41.

McCrae, R. R., and John, O. P. (1992). An Introduction to the Five-factor Model and Its Applications. *Journal of Personality*, *60*(2), 175–15.

McCullough, M. E., Emmons, R. A., and Tsang, J.A. (2002). The Grateful Disposition: A Conceptual and Empirical Topography. *Journal of Personality and Social Psychology, 82*(1), 112–27.

McCullough, M. E., Pargament, K. I., and Thoresen, C. E. (2000). *Forgiveness: Theory, Researchand Practice.* New York: Guilford Press.

McCullough, M. E., Rachal, K. C., Sandage, S. J., Worthington, E. L., Jr., Brown, S. W., and Hight, T. L. (1998). Interpersonal Forgiving in Close Relationships. II: Theoretical Elaboration and Measurement. *Journal of Personality and Social Psychology*, *75*(6), 1586–03.

McCullough, M. E., and Snyder, C. R. (2000). Classical Sources of Human Strength: Revisiting an Old Home and Building a New One. *Journal of Social and Personality Psychology, 19*(1), 1–10

McCullough, M. E., and Witvliet, C. V. (2002). The Psychology of Forgiveness. In C. R. Snyder and S. J. Lopez (Eds.), *Handbook of Positive Psychology* (pp. 446–58). New York: Oxford University Press.

Molloy, M. (2010). *Experiencing the World's Religions. Tradition, Challenge and Change. (5th.ed).* New York, NY: McGraw-Hill.

Mullet, E., Girard, M., and Bakhshi, P. (2004). Conceptualizations of Forgiveness. *European Psychologist*, *9*(2), 78–86.

Nelson, D. L., and Cooper, C. L. (2007). *Positive Organizational Behavior*. London: Sage Publications.

Nicholson, I. A. M. (1998). Gordon Allport, Character, and the Culture of Personality, 1897–937. *History of Psychology, 1*(1), 52–68.

Nuyen, A. (2009). Moral Obligation and Moral Motivation in Confucian Role-Based Ethics. *Dao: A Journal of Comparative Philosophy, 8*(1), 1–11.

O'Byrne, K. K., Lopez, S. L., and Peterson, S. (2000). *Building a Theory of Courage: A Precursor to Change?* Paper Presented at the 108th Annual Meeting of the American Psychological Association, Washington, D. C.

Paranjpe, A. C., and Rao K. R. (2008). Psychology in the Advaita Vedanta. In K. R. Rao, A. C. Paranjpe and A. K. Dalal (Eds.), *Handbook of Indian Psychology* (pp. 253–285). New Delhi: Cambridge University Press.

Park, N., and Peterson, C. (2005). The Values in Action Inventory of Character Strengths for Youth. In K. A. Moore and L. H. Lippman (Eds.), *What Do Children Need to Fourish: Conceptualizing and Measuring Indicators of Positive Development* (pp. 13–23). New York: Springer.

Park, N., and Peterson, C. (2006). Moral Competence and Character Strengths among Adolescents: The Development and Validation of the Values in Action Inventory of Strengths for Youth. *Journal of Adolescence, 29*(6), 891–909.

Park, N., and Peterson, C. (2008). Positive Psychology and Character Strengths: Application to Strengths Based School Counseling. *Professional School Counseling, 12*(2), 85–92.

Park, N., Peterson, C., and Seligman, M. E. P. (2004). Strengths of Character and Well-being. *Journal of Social and Clinical Psychology, 23*(5), 603–19.

Park, N., Peterson, C., and Seligman, M. E. P. (2006). Character Strengths in Fifty-four Nations and the Fifty US States. *Journal of Positive Psychology, 1*(3), 118–29.

Peterson, C., and Park, N. (2004). Classification and Measurement of Character Strengths: Implications for Practice. In P. A. Linley and S. Joseph (Eds.), *Positive Psychology in Practice* (433–46). Hoboken, NJ: John Wiley & Sons Inc.

Peterson, C., and Park, N. (2006). Character Strengths in Organizations. *Journal of Organizational Behavior, 27*(8), 1149–54.

Peterson, C., Park, N., Pole, N., D'Andrea, W., and Seligman, M. E. P. (2008). Strengths of Character and Post-traumatic Growth. *Journal of Traumatic Stress, 21*(2), 214–17.

Peterson, C., and Seligman, M. E. (2003). Character Strengths before and after September 11. *Psychological Science, 14*(4), 381-84.

Peterson, C., and Seligman, M. E. P. (2004). *Character Strengths and Virtues: A Handbook and Classification*. Washington, D. C.: American Psychological Association.

Puligandla, R. (1985). *Jñāna-Yoga—The Way of Knowledge (An Analytical Interpretation)*. New York: University Press of America.

Pury, C. L., and Kowalski, R. M. (2007). Human Strengths, Courageous Actions, and General and Personal Courage. *Journal of Positive Psychology, 2*(2), 120–28.

Romero, C., Friedman, L. C., Kalidas, M., Elledge, R., Chang, J., and Liscum, K. R. (2006). Self-Forgiveness, Spirituality, and Psychological Adjustment in Women with Breast Cancer. *Journal of Behavioral Medicine, 29*(1), 29–36.

Ross, L., and Nisbett, R. E. (1991). *The Person and the Situation.* Philadelphia, PA: Temple University Press.

Rushton, J. P. (1984). The Altruistic Personality. In E. Straub, D. Bar Tal, J. Karylowski, and J. Reykowski (Eds.), *Development and Maintenance of Prosocial Behavior* (pp. 271–90). New York, NY: Praeger.

Ryff, C. D., and Keyes, C. L. M. (1995). The Structure of Psychological Well-being Revisited. *Journal of Personality and Social Psychology, 57*(6), 1069–81.

Sangharakshita (1991). *The Three Jewels: An Introduction to Buddhism*. Glasgow, UK: Windhorse.

Seligman, M. E. P., and Peterson, C. (2003). Positive Clinical Psychology. In L. G. Aspinwall and U. M. Staudinger (Eds.), *A Psychology of Human Strengths: Fundamental Questions and Future Directions for a Positive Psychology* (pp. 305–17). Washington, DC: American Psychological Association.

Seligman, M. E. P., Steen, T. A., Park, N., and Peterson, C. (2005). Positive Psychology Practice: Empirical Validation of Interventions. *American Psychologist, 60*(5), 410–21.

Shryack, J., Steger, M., Krueger, R., and Kallie, C. (2010). The Structure of Virtue: An Empirical Investigation of the Dimensionality of the Virtues in Action Inventory of Strengths. *Personality and Individual Differences, 48*(6), 714–19.

Shun, W. L. (1997). *Menicus and Early Chinese Thought*. Stanford: Stanford University Press.

Sinha, D. M. (2005, January 21). The Act of Giving Spontaneously. The Speaking Tree. *Times of India.*

Slife, B. D. (2012). *Virtue Ethics in Practice: The Greenbrier Academy. Journal of Theoretical and Philosophical Psychology, 32*(1), 35–42.

Slife, B. D., and Richardson, F. C. (2008). Problematic Ontological Underpinnings of Positive Psychology: A Strong Relational Alternative. *Theory and Psychology, 18*(5), 699–23.

Snyder, C. R., and Lopez, J. L. (2007). *Positive Psychology: The Scientific and Practical Explorations of Human Strengths*. New Delhi: Sage.

Stalker, J. (1902). *The Seven Cardinal Virtues*. New York: American Tract Society.

Staudinger, U. M., Lopez, D. F., and Baltes, P. B. (1997). The Psychometric Location of Wisdom Related Performance: Intelligence, Personality, and More? *Personality and Social Psychology Bulletin, 23*(11), 1200–14.

Steen, T., Kachorek, L. V., and Peterson, C. (2003). Character Strengths Among Youth. *Journal of Youth and Adolescence, 32*(1), 5–16.

Steger, M. F., Hicks, B. M., Kashdan, T. B., Krueger, R. F., and Bouchard, T. J. Jr, (2007). Genetic and Environmental Influences on the Positive Traits of the Values in Action Classification, and Biometric Covariance with Normal Personality. *Journal of Research in Personality, 41*(3), 524–39.

Sternberg, R. (1985). Implicit Theories of Intelligence, Creativity, and Wisdom. *Journal of Personality and Social Psychology, 49*(3), 607–27.

Sussman, W. I. (1973). *Culture as History.* New York, NY: Pantheon.

Tjeltveit, A. C. (2003). Implicit Virtues, Divergent Goods, Multiple Communities: Explicitly Addressing Virtues in the Behavioral Sciences. *American Behavioral Scientist, 47*(4), 395–14.

Tsang, J., McCullough, M. E., and Fincham, F. (2006). The Longitudinal Association Between Forgiveness and Relationship Closeness and Commitment. *Journal of Social and Clinical Psychology, 25*(4), 448–72.

Van Eeden, C., Wissing, M. P., Dreyer, J., Park, N., and Peterson, C. (2008). Validation of the Values in Action Inventory of Strengths for Youth Among South African Learners. *Journal of Psychology in Africa, 18*(1), 143–54.

van Norden, B. (2007). *Virtue Ethics and Consequentialism in Early Chinese Philosophy*. New York: Cambridge University Press.

Walker, L. J., and Pitts, R. C. (1998). Naturalistic Conceptions of Moral Maturity. *Developmental Psychology, 34*(3), 403–19.

Wetlesen, J. (2002). Did Santideva Destroy the Bodhisattva Path? *Journal of Buddhist Ethics*, 9(1), 34–88.

Witvliet, C. V. O., Ludwig, T. E., and Vander Laan, K. L. (2001). Granting Forgiveness or Harboring Grudges: Implications for Emotion, Physiology, and Health. *Psychological Science, 12*(2), 117–23.

Worthington, E. L. Jr. (2006). *Forgiveness and Reconciliation: Theory and Application.* New York: Brunner–Routledge.

Xunzi (2001). 'Xunzi' in EL Hutton (tr), P. J. Ivanhoe and B. W. van Norden (Eds.), *Readings in Classical Chinese Philosophy*. New York: Seven Bridges.

CHAPTER 4

Character Strengths and Health

Amrita Yadava

Professor, Department of Psychology, Maharishi Dayanand University, Rohtak

Abstract: The concept of virtues, which are the core characteristics of an individual, and character strengths are positive traits reflected in thoughts, feelings and behaviour and has been proposed by positive psychologists to provide a description and classification of traits, emotions and beliefs that enable humans to thrive. Common sense connotes the role of strength of character in coping with adverse circumstances. This strength of character is generally considered to be 'moral' rather than 'physical' as physical strength is a natural endowment and moral strength is acquired through initiative and effort. This moral strength is connoted as a virtue only when it is exercised to overcome obstacles. A number of studies show that character strengths are associated with better quality of life, faster recovery and adherence to medical regime and negatively to depression and mortality. Elevations of some specific character strengths have been reported to help in coping with stressful situations and leading to positive consequences, thereby providing support to the modulator role of character strengths in health and well-being. Positive mood, optimistic outlook, social support and happiness have been found to be associated with better health and longevity, indicating that positive traits might act as buffers against illness and promote psychological well-being. Researchers have proposed a strength-based approach to improving job satisfaction which involved identification of strengths associated with a particular job, integration of these strengths into the worker's self-image, such that he/she defined himself/herself according to the signature themes, and lastly attribution of success by worker to these signature strengths, thereby leading to satisfaction and productivity. Seligman, Steen, Park and Peterson (2005) proposed that a similar strength-based approach could be applied in the area of positive health, whereby identification of signature strengths associated with health and psychological well-being could lead to the development of intervention programmes as use of signature strengths leads to positive emotions and well-being both in the present as well as in future. The present article focuses on the linkage of character strengths with health and psychological well-being.

Character Strengths and Health

Generally, ill health is equated with disease and infirmity and is considered as the antithesis of health, while being healthy is a coveted state. In 1973, Rokeach, while developing his value survey, had originally planned to use health as one of the values. However, there was no variability with regard to health as everybody ranked health first. This indicated that nearly everyone valued health more than any other personal asset. Paradoxically, the behaviour of people does not reflect this, as they do not behave in ways that maximize health and minimize disease and disability. Common sense connotes the role of strength in coping with disease and other life stresses. This strength is generally considered to be 'moral' rather than 'physical' as physical strength is perceived as a natural endowment, and moral strength is acquired through initiative and effort. This moral strength is connoted as a value or virtue only when it is exercised to overcome obstacles.

Values are social agreements about what is right, good and to be cherished, which contain cognitive and affective elements, and therefore, act as implicit or explicit guides for action and framing what is sought after and what is to be avoided. People appeal to these social or moral beliefs as the ultimate rationale for their action and they provide a means of self-regulation of impulses as these are individually endorsed and highly accessible to the individual so that individuals can maintain harmony with the group within which they live. Values predict an individual's behaviour, yet at times, even personally endorsed values do not influence action unless they are made salient to the individual at the time of action. Moreover, in any given situation, more than one personally endorsed value may apply, and the behavioural choice appropriate for one value may conflict with the behavioural choice appropriate to another value (Oyserman, 2002). Seligman and Peterson (2000) advocated development of the strengths which would be instrumental in dealing with the present stressors and equip the individual with the ability to mature, function well, become successful, find personal fulfilment and optimum potential. They developed the concept of values in action; to identify the internally held beliefs (and their levels of behaviours) which shape a person's destiny. These values were operationally defined as positive traits of an individual which were classified as character strengths and virtues (CSV) to provide a description and classification of traits that enable human thriving along the same lines as the Diagnostic and Statistical Manual of Mental Disorders (DSM) of the American Psychological Association (APA) does for psychological disorders that disable human beings (Peterson and Seligman, 2004). The present chapter focuses on the relationship between character strengths and health.

Health–Behaviour Linkage

Health is a tangible yet elusive concept. It is an implicit characteristic of an individual which is described as a feeling of equilibrium or well-being, although it is perceived only when there is an imbalance or reduction. On the other hand, illness is acquired and is perceived by its presence. The advent of positive psychology at the beginning of the present millennium culminated in the conceptualization of the concept of *positive health*, where health was considered as a positive state, much beyond the mere absence of disease, which was operationalized by a combination of excellent status on biological, subjective and functional measures and suggested that a focus on health rather than illness would be cost saving and lifesaving (Seligman, 2008). The positive view of health implicates the linkage of physical, psychological and social variables to longevity, where being healthy is conceived as a state that can be achieved only by continuous effort. A person who does not consciously exert effort to maintain his/her health cannot be healthy, even if there is nothing functionally or biologically wrong with the person.

Various health-behaviour models (Health Belief Model; Theory of Reasoned-Action; Self-Regulation Theory; Precaution Adoption Process Model) have been proposed which implicate a role of knowledge, attitudes and beliefs in health related behaviour, as the individual's perception about the disease and his own ability to cope with the disease strongly affect the ability to manage diseases and have an impact on every domain of health. The relationship of attitude, belief and values with health has been demonstrated by a number of researchers. Brownlee–Deffeck, Peterson, Simons, Goldstein, Kilo and Hoette (1987), examined health beliefs in diabetic patients and reported a facilitative effect of these beliefs on adherence to treatment regimen. Domingo (1989) empirically demonstrated that the Theory of Reasoned Action predicted beliefs about the behaviour, as well as beliefs about the consequences of performing the behaviour.

Illness intrusiveness also has a direct influence on health behaviours when the condition interferes with valued activities and interest, and indirectly through reduced perception of personal control, self-efficacy and self-esteem. Penninx, Tilburg, Boeke, Deeg, Kriegsman and Eijk (1998) have shown that having strong social support and good personal coping resources, along with high self-esteem, a sense of mastery and feelings of self-efficacy in the face of adversity are associated with fewer depressive symptoms in people with diabetes. For adults, the belief that compliance would benefit health predicted adherence while among adolescents adherence was influenced by perceived severity of illness and the belief that they were susceptible to diabetic complications if the treatment regimen was not followed.

Subjective norms, which are internalized beliefs, are learnt through modelling and observational learning and, are generally, in cognizance with the cultural norms. These, in turn, assume control of the individual's behaviour and can be equated with values of the individual. Nyazema (1984) found that diabetic and hypertensive patients in Zimbabwe who believed in traditional healers did not adhere to modern medical procedures. Ruiz and Ruiz (1983) reported that Latino patients readily complied with the recommendation of physicians who demonstrated understanding of Hispanics cultural norms and practices. Finlay, Trafimow and Moroi (1999) reported that people whose behaviours are generally under normative control have the intention to perform more healthy behaviours than the people whose behaviours are generally under attitudinal control. Taken together, these researches provide support for the fact that attitude, beliefs and values influence health behaviour and recovery from illness.

Virtues and Character Strength

Values can be conceptualized as codes or general criteria which represent what is expected and hoped for and what is avoided or forbidden, which play a decisive role in the occurrence of behaviour. Proponents of positive psychology have proposed that good character is a function of virtues which are the core characteristics of the individual and character strengths are positive traits reflected in thought, feeling and behaviour, which exist in degrees rather than all or none categories and can be measured as individual differences. These virtues and character strengths are ubiquitous and probably universal as they are endorsed by almost every culture across the world (Dahlsgard, Peterson and Seligman, 2005). Peterson and Seligman (2004) speculate that these virtues are grounded in biology through an evolutionary process that led to their selection as a means of managing the important tasks necessary for the survival of the species.

Peterson and Seligman (2004) developed the Values in Action (VIA) classification of strengths as an initial step towards specifying the important positive traits.The strengths were identified via an esoteric route by analyzing the religious teachings of major Eastern (Taoism, Confucianism, Hinduism and Buddhism) and Western (Judaism, Christianity, Athenian virtue and Islam) religion and going into authoritative texts of these religions. They identified twenty-four character strengths which were further grouped under six virtues (wisdom and knowledge, courage, humanity, justice, temperance and transcendence). Signature strengths (top five character strengths) have been found to vary with demographic variables, such as age, gender and culture (Linley *et al.*, 2007; Priti, 2010; Singh and Choubisa, 2010). However, the strength of fairness was universally endorsed.

The classification of character strengths into virtues was purely theological as empirical research has not revealed six dimensions of character strengths. Some researchers have identified four or five dimensions (Park and Peterson, 2005, 2006; Peterson, Park, Pole, D'Andrea, and Seligman, 2008; Priti, Yadava and Sharma, 2011a). A factor analysis of VIA by Brdar, Todd and Kashdan (2010) revealed a four-factor solution (interpersonal strengths, fortitude, vitality and cautiousness) where vitality (with zest, hope, curiosity and humour as indicators) emerged as most relevant to well-being. When examining individual strengths, zest, curiosity, gratitude and optimism/hope emerged with the strongest associations with elevated life satisfaction, subjective vitality, satisfaction of autonomy, relatedness and competence needs, and a pleasurable, engaging and meaningful existence. Park, Peterson and Seligman (2004) had also reported that strengths 'of the heart'—zest, gratitude, hope and love—were more robustly associated with life satisfaction than cerebral strengths, such as curiosity and love of learning.

On the other hand, Van Eeden, Wissing, Dreyer, Park and Peterson (2008) and Singh and Choubisa (2010) found that all the twenty-four strengths formed a single cluster. The emergence of a single factor solution has been interpreted in terms of a master value which might be responsible for covertly controlling the expression of other character strengths deliberatively when called into action. A study by Peterson and Seligman (2003) provides support for the above view. Comparison of VIA scores of 4817 American respondents, two months after 9/11 with scores of those individuals who had completed the survey before September 11, revealed that seven character strengths, i.e., gratitude, hope, kindness, leadership, love, spirituality and teamwork were higher in the later assessed sample. These strengths were still elevated when assessed two months later although their degree was somewhat lesser than immediately following the attacks. Thus, it appears that elevations of some specific character strengths help in coping with stressful situations and lead to positive consequences.

Character Strengths and Health

Positive psychology postulates two general approaches to happiness and good life: (1) the eudaemonist approach, which focuses on doing and thinking rather than emotions (virtue, character and morality); and (2) the hedonic approach, where emphasis is on achieving a balance between positive and negative affect (finding pleasure or happiness and avoiding pain). Positive health which emerged from positive psychology suggests that positive affective states and purposeful life experiences influence the immune system and lead to psychological well-being. Further, it advocates that a focus on health rather than illness would be more economical both in terms of saving cost as well as life. It can be operationalized by a combination of excellent status on biological, subjective and functional measures. Positive health predicts increased longevity, decreased health costs, better mental health in aging and better prognosis when illness strikes (Seligman, 2008).

Empirical evidence implicates a role of character strengths in maintenance of health. Qualitative differences in character strengths have been reported in healthy and unhealthy groups. Comparison of character strengths between healthy and unhealthy (non-disease, but reporting lower score on perceived health measures) respondents revealed significant differences in only one character strength, i.e., self-regulation (Priti, 2010). Further, spirituality surfaced as the topmost signature strength of the unhealthy group while it did not even figure in the top five strengths of the healthy sample, i.e., open mindedness, love, teamwork, self-regulation and humour (Priti, Yadava and Sharma, 2011b). Proyer, Gander, Wellenzohn and Ruch (2013) reported greater endorsement of character strengths with a number of health behaviours and physical fitness. All character strengths (except humility and spirituality) were associated with multiple health behaviours. While self-regulation had the highest overall association, curiosity, appreciation of beauty/excellence, gratitude, hope and humour also displayed strong connections with health behaviours. However, Eracleous (2008) failed to observe any significant differences in the character strengths of healthy and unhealthy respondents.

Perusal of the studies relating to non-diseased sample reveals a positive relationship between emotional and social intelligence and positive psychological health (Hooda, Sharma and Yadava, 2008; Hooda, Sharma and Yadava, 2009). Govindji and Linley (2007) and Proctor, Maltby and Linley (2009) in studies on university students (214 and 135 students, respectively), showed that people who used their strengths more, reported higher levels of subjective well-being (i.e., happiness) and psychological well-being (i.e., fulfilment). Park, Peterson and Seligman (2004) reported that life satisfaction had a robust correlation with hope, zest, gratitude, love and curiosity. In contrast, modesty and the intellectual strengths were weakly associated with life satisfaction. A monotonic relationship was observed between character strengths and life satisfaction indicating that excess of any strength does not diminish life satisfaction. Results of an internet based intervention study (Seligman, Park and Peterson, 2005) indicate that participants tended to be happier and less depressed immediately post test (one week after intervention) and at every testing (up to six months). These results indicate that identification of character strengths led to positive effect on the immediate post test, while using signature strengths in new ways lead to steady improvement up to one month, which lasted ever after six months. Harker and Keltner (2001) reported that women who flashed a genuine smile in their yearbook photos as freshmen have more marital satisfaction twenty-five years later. Optimism in early adulthood has been found to be a good predictor of health in later adulthood over periods up to thirty-five years (Peterson, Seligman and Valliant, 1998). Peterson (2000) reported that optimistic people actually avoid stress in lives by forming better social support networks, having healthier life styles and a better immune system which shields them from the development of illness. In looking at more severe physiological events, positive affect and positive explanatory styles have been found to be protective against stroke (Ostir, Markides, Peek and Goodwin, 2001), rapid progression of HIV (Taylor, Kemeny, Reed, Bower, and Gruenewald, 2000), and general mortality rates in the elderly (Cohen and Pressman, 2006; Maruta, Colligan, Malinchoc and Offord, 2000). Study of a sample of 334 Swiss adults and 634 peer (informant) ratings, suggested that hope, zest and curiosity (and gratitude and love) have key roles in the connection between character strengths and life satisfaction. Informant reports also related positively to the endorsement of pleasure, engagement, and meaning (Buschor, Proyer and Ruch, 2013).

Positive emotions (hope, optimism, etc.) and humour reduce the risk of infection, such as common cold (Cohen, Alper, Doyle, Treanor and Turner, 2006). Fredrickson (2003) found that positive emotions undo the cardiovascular effects of negative emotions when people experience stress. If individuals do not regulate the physiological changes once the stress is past, they can lead to illness, such as coronary heart disease (CHD) and heightened mortality. Study of children within the age range of three–nine years showed that love, zest and hope enhanced happiness in younger children while gratitude led to happiness in older children (Park and Peterson, 2006). Forgiveness has also been found to produce beneficial effects directly by reducing the physiological load associated with psychological stress (betrayal and conflict) and indirectly through reduction in perceived stress (Lawyer *et al.*, 2004).

Empirical evidence also provides support to the modulatory role of character strengths in health maintenance and coping with diseases. Spirituality has been found to be related to quality of life and is an effective coping mechanism (Leung, Chiu and Chen, 2006; Riley, Perna, Tate, Forchheimer, Anderson and Luera, 1998; Tuck, McCain and Elswick, 2001). It lowers fear of death and discomfort and improves emotional adjustment by reducing health compromising behaviours (Hafen, Karren, Frandsen and Smith, 1996), thereby helping in adaptation to chronic illness (Livneh, Lott and Antonak, 2004).

Priti, Yadava and Sharma (2011a) reported a qualitative difference in factor structure of VIA in diseased (arthritis and diabetic patients) and non-diseased respondents, which implicated a role of fostering gratitude and spirituality in arthritis patients and curiosity and creativity in diabetics in coping with the disease and improving prognosis. Comparison of the signature strengths of the non-diseased and chronic disease groups showed that the strength of fairness was common across the signature strengths of all the three samples (non-diseased and two chronic diseases). In both the disease groups, spirituality was found to have the maximum score while it did not even figure in the signature strengths of the non-diseased sample (Priti, 2010), thereby providing support for the fact that elevations of some specific character strengths help in coping with a stressful situation and leads to positive consequences.

Since optimism reduces the risk of depression, it leads to better overall health and strong emotional coping skills, acts as a protective factor (Giltay, Geleijnse, Zitman, Hoekstra, and Schouten, 2004; Kubazansky, Sparrow, Vokonas and Kawachi, 2001; Livneh, Lott and Antonak, 2004) and facilitates recovery (Scheier *et al.*, 1989) by improving positive aspects of recovery (Chamberlain, Petrie and Azariah, 1992), such as greater adherence to medical regime (Leednam, Meyerowitz, Muirhead and Frist, 1995), thereby reducing mortality (Buchanan, 1995; Giltay *et al.*, 2004; Ostir, Ottenbacher and Markides, 2004).

Humour helps in successful adaptation to live with chronic illness (Peterson, Park and Seligman, 2006b). Forgiveness also leads to better recovery as it helps to reduce the psychological load by reducing the perceived stress (Lawyer *et al.*, 2004). Carver, Scheier and Pogo (1992) reported that optimists adapted better after surgery and appeared to get along with their normal lives at a more rapid pace. Further, optimism was found to be predictive of physical and mental health, which in turn, was found to be mediated by positive mood, immunological robustness, effective coping and health promoting behaviour.

People who believe that they are personally responsible for their personal health are more likely to adhere to medical advice (Heiby, Gafaria and McCann, 1989). The importance of sense of control extends to the beliefs about health outcomes determined by powerful people in the environment, such as health care providers. For patients, who had severe rheumatoid arthritis, beliefs in internal control have been found to be associated with greater mood disturbance. It might be argued that it is bothersome to have exacerbations of a serious illness when one thinks that these events are under their control (Affleck, Tennen, Pfeiffer and Fifeld, 1987). Sherbourne, Hays, Ordway, Dimatteo and Kravitz (1992) reported that patients who use avoidance coping strategies like denying personal vulnerability or avoiding personal responsibility for taking actions are less likely to comply with the doctor's advice.

With regard to specific diseases, character strengths have been mainly studied in chronically ill patients (cancer, CHD, HIV). People who express positive emotions come down with fewer colds and flues after being exposed to the viruses than those who express negative emotions like anger, sadness or stress (Cohen and Pressman, 2006). In patients with CHD and heart ailment, optimism and spirituality were found to reduce mortality (Giltay *et al.*, 2004) as it protected against cardiovascular events (Kubazansky *et al.*, 2001), improved adherence to medical advice (Leednam *et al.* ,1995) and increased the speed of recovery from depression in the chronically ill patients (Durham, 1998). Patients who show a fighting spirit seem to do better than those who do not. Patients with higher ratings of hopelessness and negative beliefs about the future had poorer health outcomes (Goodkin, Antoni and Balaney, 1986).

Pettingale, Morris, Greer and Haybittle (1985) have also reported that patients with a fighting spirit are more likely to survive breast cancer.

In cancer patients, spirituality and humour were found to be an effective coping mechanism (Ferrell, Smith, Juarez and Melancon, 2003). Religious activities and spirituality reduced the rates of cancer (Hafen *et al.*, 1996) and distress (Laubmeier, Zakoswski and Bair, 2004) leading to better quality of life and increased life satisfaction (Riley, Perna, Tate, Forchheimer, Anderson and Luera, 1998). Appreciation of beauty, bravery, curiosity, fairness, forgiveness, humour, gratitude, love of learning and spirituality was higher in the diseased respondents (Peterson, Park and Seligman, 2006a). However, comparison of character strengths of adolescents who have survived cancer to healthy adolescents by Guse and Eracleous (2011) revealed that the experience of serious illness, such as cancer neither hindered nor enhanced the development of character strengths in the group of adolescent survivors. Comparison of VIA scores of cancer patients with healthy respondents indicated higher scores on humour, fairness, zest, honesty and perspective (Peterson, Park and Seligman, 2006a) although another study (Eracleous, 2008) reported no differences.

In HIV patients, spirituality was found to be positively associated with well-being (Debra, and Elaine, 1998) as it was related with quality of life, social support, and negatively with perceived stress, uncertainty and psychological distress (Tuck, McCain and Elswick, 2001). Life satisfaction in HIV patients was found to be associated with the character strengths of hope, zest, gratitude, love of learning and curiosity (Park, *et al.*, 2004).

In orthopedic patients (amputation, spinal injury), non-spiritual patients were found to have lower quality of life and life satisfaction (Riley, *et al.*, 1998). An in-depth interview of osteoarthritis patients revealed five major themes, i.e., strengths, prudence, gratitude, self worth and insight into flourishing (Swift, Ashcroft, Todd, Campbell and Dieppe, 2002).

Although intervention studies related to enhancement of character strength are sparse, they indicate that use of signature strengths lead to positive emotions, well-being, coping as well as recovery from disease. Proyer, Ruch and Buschor (2013) compared the impact of character strengths-based positive interventions in adults. Pre-post comparison of life satisfaction of a group trained on strengths of curiosity, gratitude, hope, humour and zest with a group trained on strengths that usually demonstrate low correlations with life satisfaction, i.e., appreciation of beauty and excellence, creativity, kindness, love of learning, and perspective and a wait-list control group indicated that strengths that correlate highly with life satisfaction should be addressed in strengths-based interventions. Subjective ratings of well-being after the interventions concluded, indicated gains in both the intervention groups above that of a wait-listed control group, thereby implicating the potential of strength-based interventions for improving well-being. However, analyses underscore the special role of self regulation in facilitating success in the interventions.

Random assignment to a group instructed to use two signature strengths or use one signature strength and one bottom strength revealed significant gains in satisfaction with life compared with a control group but no differences between the two treatment groups (Rust, Diessner and

Reade, 2009). The identification of signature strengths followed by discussion with a friend about strengths and use of three signature strengths in daily life was reported to boost cognitive (but not affective) well-being at three months' follow-up (Mitchell, Stanimirovic, Klein and Vella-Brodrick, 2009).

Peterson, Park and Seligman (2006a) conducted a retrospective web-based study of 2087 adults and reported small but reliable associations between a history of physical illness and the character strengths of appreciation of beauty, bravery, curiosity, fairness, forgiveness, gratitude, humour, kindness, love of learning and spirituality. A history of psychological disorder and the character strengths of appreciation of beauty, creativity, curiosity, gratitude and love of learning were also associated. A history of problems was linked to decreased life satisfaction, but only among those who had not recovered. In the case of physical illness, less of a toll on life satisfaction was found among those with the character strengths of bravery, kindness and humour, and in the case of psychological disorder, less of a toll on life satisfaction was found among those with the character strengths of appreciation of beauty and love of learning. They suggested that recovery from illness and disorder may benefit character.

Thus, it appears that understanding the relationship between stress, health and illness, identifying risk factors for sickness and protective factors for health, developing programmes for illness prevention and for ensuring compliance with medical treatment regimens (Blumenthal, Matthews and Weiss, 1994) can be furthered by understanding the links between character strengths and health. Although major differences in character strengths have not been observed in healthy and unhealthy groups, self-regulation has been found to be associated to health in healthy respondents. Surprisingly, it appears to lose its relevance once disease sets in. The reason could be that other strengths gain more importance as they are required to deal with the stresses associated with the disease or may be these strengths may increase as the perspective of the individual changes as a consequence of the disease, where things which were taken for granted earlier, now assume greater significance.

The above reviewed studies clearly reflect that positive affect and traits have an impact on health enhancing behaviours, act as buffers against illness, enhance physiological functioning as well as compliance with medical treatment regimes. Specific character strengths (mainly the strengths of the heart) have been found to be associated with health maintenance and prognosis from chronic diseases. Thus, interventions which increase positive affect (happiness, optimism, hope, flow) could be helpful in reducing the stress associated with the disease, boost the immune system and facilitate adherence to medical regime. Further, strengths of the heart (spirituality, forgiveness, gratitude) may go a long way in helping the patient to focus on intrinsic goals (emotional intimacy, community service, personal growth) and accept the disease as inevitable. This could set the stage for positive change as acceptance and giving meaning to the present reality is necessary to let go of the negative emotions and stresses associated with one's limitations along with existential anxiety, make the most of one's brief life by living authentically and passionately, find recourse in faith, meaning and relationships, thereby accepting what cannot be changed. Wong (2012) developed a resource-congruence model of effective coping (Meaning Management

Theory), where acceptance is implicated as a strategy for coping with problems and stresses that are appraised as uncontrollable where meaning management refers to how one manages meaning-seeking, meaning-making and meaning reconstruction in order to survive and flourish. People are considered as meaning-seeking and meaning-making a creature that live in a world of meaning, attribute meaning to events that affect them more than the events themselves, and offers them the best protection against threats and losses and the best means to achieve well-being. Thus, acceptance can be an effective and versatile way of adaptation, capable of repairing the worst and bringing out the best in a person. Wong has suggested five pathways of acceptance which cover five major life domains—personhood; relationships; existential issues regarding the human condition; stress, loss and suffering; and mindfulness. This is presently an important, but under-researched area in the positive psychology of well-being which appears to have a lot of potential in health maintenance and prognosis in chronic diseases.

Thus, it can be concluded that a conflux of some character strengths might help in dealing with chronic disorders because onset of these disorders demands a major change in the lifestyle or perspective of the individual. Since maintaining health and dealing with disease, specially chronic disorders which have a prolonged span or are not treatable, require a diligent health/treatment regime focusing on health promoting behaviours, reducing the disease symptoms and/or slowing down the progression of diseases, character strengths, such as patience, perseverance, self-regulation, hope, spirituality, forgiveness, gratitude, creativity, etc., could go a long way in dealing with stress-associated health and disease. A strength-based approach, as proposed by Seligman, Steen, Park and Peterson (2005) could be applied in the area of positive health, whereby identification of signature strengths associated with health and prevention/prognosis of diseases could lead to the development of intervention programmes for managing chronic disorders as use of signature strengths leads to positive emotions and well-being both in the present as well as future.

References

Affleck, G., Tennen, H., Pfeiffer, C., and Fifeld, J. (1987). Appraisals or Control and Predictability in Adapting to a Chronic Disease. *Journal of Personality and Social Psychology, 53*(2), 273–79.

Blumenthal, S. J., Matthews, K., and Weiss, S. M. (1994). New Research Frontiers and Behavioral Medicine: *Proceedings of the National Conference.* (NIH Publication No. 94–3772). Washington DC: U.S. Printing Office.

Brdar, I., Todd B., and Kashdan, T. B. (2010). Character Strengths and Well-being in Croatia: An Empirical Investigation of Structure and Correlates. *Journal of Research in Personality, 44*(1), 151–54.

Brownlee-Deffeck, M., Peterson, L., Simond, J. F., Goldstein, D., Kilo, C., and Hoette, S. (1987). The Rate of Health Beliefs in the Regimen Adherence and Metabolic Control of Adolescents and Metabolic Control of Adolescents and Adults with Diabetes Mellitus. *Journal of Consulting and Clinical Psychology, 55*(2), 139–44.

Buchanan, G. M. (1995). Explanatory Style and Coronary Heart Disease. In G. M. Buchanan and M. E. P. Seligman (Eds.), *Explanatory Style* (pp. 225–32). Hillsdale, NJ: Erlbaum.

Buschor, C., Proyer, R. T., and Ruch, W. (2013). Self-and Peer-rated Character Strengths: How Do they Relate to Satisfaction with Life and Orientations to Happiness? *Journal of Positive Psychology, 8*(2), 116–27.

Carver, C. S., Scheier, M. F., and Pogo, C. (1992). Conceptualizing the Process of Coping with Health Problems. In H. S. Friedman (Ed.), *Hostility, Coping, and Health* (pp. 167–87). Washington DC: American Psychological Association.

Chamberlain, K., Petrie, K., and Azariah, R. (1992). The Role of Optimism and Sense of Coherence in Predicting Recovery Following Surgery. *American Journal of Psychology and Health, 7*(4), 301–10.

Cohen, S., and Pressman, S. D. (2006). Positive Affect and Health. *Current Directions in Psychological Science, 15*(3), 122–25.

Cohen, S., Alper, C. M., Doyle, W. J., Treanor, J. J. and Turner, R. B. (2006). Positive Emotional Style Predicts Resistance to Illness After Experimental Exposure to Rhinovirus or Influenza a Virus. *Psychosomatic Medicine, 68*(6), 809–15.

Dahlsgaard, K., Peterson, C., and Seligman, M. (2005). Shared Virtue: The Convergence of Valued Human Strengths across Culture and History. *Review of General Psychology, 9*(3), 203–13.

Debra, P. O. N., and Elaine K. K. (1998). Spirituality and Chronic Illness. *Journal of Nursing Scholarship, 30*(3), 275–80.

Domingo, M. A. B. (1989). *Attitude-Behavior Relationship: A Comparative Analysis Using the Health Belief Model and the Theory of Reasoned Action*, Unpublished M. A. Thesis, University of Delaware, USA.

Durham, N. C. (1998). Scientists Find Direct Link between Religious Faith and Recovery from Depression. *American Journal of Psychiatry, 155*(4), 536–42.

Eracleous, G. (2008). *Character Strengths of Adolescents Who Have Survived Cancer: A Comparative Study.* Ph. D. Thesis, University of Johannesburg.

Ferrell, B. R., Smith, S. L., Juarez, G., and Melancon, C. (2003). Meaning of Illness and Spirituality in Ovarian Cancer Survivors. *Journal of Oncology Nursing Forum, 30*(2), 249–57.

Finlay, K. A., Trafimow, D. and Moroi, E. (1999). The Importance of Subjective Norms on Intentions to Perform Health Behaviours. *Journal of Applied Social Psychology, 29*(11), 2381–93.

Fredrickson, B. L. (2003). The Value of Positive Emotions. *American Scientist, 91*(4), 330–35.

Giltay, E. J., Geleijnse, J., Zitman, M., Hoekstra, F. G., and Schouten, E. G. (2004). Dispositional Optimism and All-cause and Cardiovascular Mortality in a Prospective Cohort of Elderly Dutch Men and Women. *Archives of General Psychiatry, 61*(11), 1126–35.

Goodkin, K., Antoni, M. and Balaney, P. H. (1986). Stress and Hopelessness in the Promotion of Cervical Intraepithelial Neoplasia to Invasive Squamous Cell Carcinoma of the Cervix. *Journal of Psychosomatic Research, 30*(1), 67–76.

Govindji, R., and Linley, P. A. (2007). Strengths Use, Self-concordance and Well-being: Implications for Strengths Coaching and Coaching Psychologists. *International Coaching Psychology Review, 2*(2), 143–53.

Guse, T., and Eracleous, G. (2011). Character Strengths of Adolescent Survivors of Childhood Cancer. *Health SA Gesondheid, 16*(1), 1–6.

Hafen, B. Q., Karren, K. J., Frandsen, K. J., and Smith, N. L. (1996). *Mind Body Health.* Boston: Allyn and Bacon.

Harker, L., and Keltner, D. (2001). Expressions of Positive Emotion in Women's College Yearbook Pictures and Their Relationship to Personality and Life Outcomes across Adulthood. *Journal of Personality and Social Psychology, 80*(1), 112–24.

Heiby, E. M., Gafaria, C. T., and McCann, S. C. (1989). Situational and Behavioural Correlates of Compliance to a Diabetic Regimen. *Journal of Compliance in Health Care, 4*(2), 101–16.

Hooda, D., Sharma, N. R., and Yadava, A. (2008). Emotional Intelligence as a Predictor of Positive Health in Working Adults. *Journal of Indian Health Psychology, 2*(2), 196–207.

Hooda, D., Sharma, N. R., and Yadava, A. (2009). Social Intelligence as a Predictor of Positive Psychological Health. *Journal of the Indian Academy of Applied Psychology, 35*(1), 143–50.

Kubazansky, L. D., Sparrow, D., Vokonas, P., and Kawachi, I. (2001). Is the Glass Half Empty or Half Full? A Prospective Study of Optimism and Coronary Heart Disease in the Normative Aging Study. *Psychosomatic Medicine, 63*(6), 910–16.

Laubmeier, K. K., Zakoswski, S. G., and Bair, J. P. (2004). The Role of Spirituality in the Psychological Adjustment to Cancer: A Test of the Transactional Model of Stress and Coping. *International Journal of Behavioral Medicine, 11*(1), 48–55.

Lawyer, K. A., Younger, J. W., Piferi, R. L., Billington, E., Jobe, R. E., Edmondson K., and Warren, J. H. (2004). A Change of Heart: Cardiovascular Correlates of Forgiveness in Response to Interpersonal Conflict. *Journal of Behavioral Medicine, 26*(5), 373–93.

Leednam, B., Meyerowitz, B. E., Muirhead, J., and Frist, W. H. (1995). Positive Expectations Predict Health after Heart Transplantation. *Health Psychology, 14*(1), 74–79.

Leung, K., Chiu, T., and Chen, C. (2006). The Influence of Awareness of Terminal Condition on Spiritual Well-being in Terminal Cancer Patients. *Journal of Pain and Symptom Management, 31*(5), 449–56.

Linley, P. A., Maltby, J., Wood, A. M., Joseph, S., Harrington, S., Peterson, C., Park, N., and Seligman, M. (2007). Character Strengths in the United Kingdom: The VIA Inventory of Strengths. *Personality and Individual Differences, 43*(2), 341–51.

Liveneh, H., Lott, S., and Antonak, R. (2004). Patterns of Psychosocial Adaptation to Chronic Illness and Disability: A Cluster Analytic Approach. *American Journal of Psychology, Health and Medicine, 9*(4), 411–30.

Maruta.T., Colligan, R. C., Malinchoc, M., and Offord, K. P. (2000). Optimists vs. Pessimists: Survival Rate Among Medical Patients over a 30-Year Period. *Mayo Clinic Proceedings, 75*(2), 140–43.

Mitchell, J., Stanimirovic, R., Klein, B., and Vella-Brodrick, D. (2009). A Randomised Controlled Trial of a Self-guided Internet Intervention Promoting Well-being. *Computers in Human Behavior, 25*(3), 749–60.

Nyazema, N. Z. (1984). Toward Better Patient Drug Compliance and Comprehension: A Challenge to Medical and Pharmaceutical Service in Zimbabwe. *Social Science and Medicine, 18*(7), 551–54.

Ostir, G. V., Markides, K. S., Peek, M. K., and Goodwin, J. S. (2001). The Association Between Emotional Well-being and The Incidence of Stroke in Older Adults. *Psychosomatic Medicine, 63*(2), 210–15.

Ostir, G. V., Ottenbacher, K. J., and Markides, K. S. (2004). Onset of Frailty in Older Adults and the Protective Role of Positive Affect. *Psychology and Aging, 19*(3), 402–08.

Oyserman, D. (2002). Values: Psychological perspectives. In N. Smelser and P. Baltes (Eds.), *International Encyclopedia of the Social and Behavioral Sciences* (Vol. 22, pp. 16150–53). New York, NY: Elsevier Science.

Park, N., and Peterson, C. (2005). The Values in Action Inventory of Character Strengths for Youth. In K. A. Moore, and L. H. Lippman (Eds.), *What do Children Need to Flourish: Conceptualizing and Measuring Indicators of Positive Development* (pp. 13–23). New York, NY: Springer.

Park, N., and Peterson, C. (2006). Moral Competence and Character Strengths among Adolescents: The Development and Validation of the Values in Action Inventory of Strengths for Youth. *Journal of Adolescent, 29*(6), 891–909.

Park, N., and Peterson, C. (2008). Positive Psychology and Character Strengths: Application to Strengths-Based School Counseling. *Professional School Counseling, 12*(2), 85–92.

Park, N., Peterson, C., and Seligman, M. (2004). Strengths of Character and Well-being. *Journal of Social and Clinical Psychology, 23*(5), 603–19.

Penninx, B. W., Van Tilburg, T., Boeke, A. J., Deeg, D. J., Kriegsman, D. M., and Van Eijk, J. T. (1998). Effects of Social Support and Personal Coping Resources on Depressive Symptoms: Different for Various Chronic Diseases. *Health Psychology*, *17*(6), 551–58.

Peterson, C. (2000). The Future of Optimism. *American Psychologist, 55*(1), 44–55.

Peterson, C, and Seligman, M. (2003). Character Strengths Before and after September 11. *Psychology Science, 14*(4), 381–84.

Peterson, C., and Seligman, M. (2004). *Character Strengths and Virtues. A Handbook and Classification.* New York: Oxford University Press.

Peterson, C., Park, N., and Seligman, M. (2006a). Strengths of Character and Well-being Among Youth. *Journal of Youth, 4*(1), 35–40.

Peterson, C., Park, N., Pole, N., D'Andrea, W. and Seligman, M. (2008). Greater Strengths of Character and Post-traumatic Growth. *Journal of Traumatic Stress, 21*(2), 214–17.

Peterson, C., Seligman, M., and Vaillant, G. E. (1998). Pessimistic Explanatory Style is a Risk Factor for Physical Illness: A Thirty-five Year Longitudinal Study. *Journal of Personality and Social Psychology, 55*(1), 23–27.

Peterson, C., Park, N., and Seligman, M. E. P. (2006b). Greater Strengths of Character and Recovery from Illness. *The Journal of Positive Psychology, 1*(1), 17–26.

Pettingale, K. W., Morris, T., Greer, S., Haybittle, J. L. (1985). Mental Attitudes to Cancer: An Additional Prognostic Factor. *The Lancet, 325*(8431), 750.

Priti (2010). *Role of Values-in-Action in Health Maintenance and Prognosis from Chronic Disease.* Unpublished Ph. D. Thesis. Department of Psychology, M. D. University, Rohtak.

Priti, Yadava, A., and Sharma, N. R. (2011a). Factor Structure of Values-in-Action in Diseased and Non-diseased Adults. *Journal of Indian Health Psychology, 5*(2), 81–90.

Priti, Yadava, A. and Sharma, N. R. (2011b). Perceived Health in Diseased and Non diseased Respondents. *Indian Journal of Psychology and Education, 1*(2), 75–86.

Proctor, C., Maltby, J., and Linley, P. A. (2009) Strengths use as a Predictor of Well-being and Health-related Quality of Life. *Journal of Happiness Studies, 10*(5), 583–30.

Proyer, R. T., Gander, F., Wellenzohn, S., and Ruch, W. (2013). What Good are Character Strengths Beyond Subjective Well-being? The Contribution of the Good Character on Self-reported Health-oriented Behaviour, Physical Fitness, and the Subjective Health Status. *Journal of Positive Psychology, 8*(3), 222–32.

Proyer, R. T., Ruch,W., and Buschor, C. (2013). Testing Strengths-Based Interventions: A Preliminary Study on the Effectiveness of a Program Targeting Curiosity, Gratitude, Hope, Humor, and Zest for Enhancing Life Satisfaction. *Journal of Happiness Studies, 14*(1), 275–92.

Riley, B., Perna, R., Tate, D., Forchheimer, M., Anderson, C., and Luera, G. (1998). Types of Spiritual Well-being Among Persons with Chronic Illness: Their Relation to Various Forms of Quality of Life. *Archives of Physical Medicine and Rehabilitation, 79*(3), 258–64.

Rokeach, M. (1973). *The Nature of Human Values.* Free Press, New York.

Ruiz, P., and Ruiz, P. P. (1983). Treatment Compliance among Hispanics. *Journal of Operational Psychiatry, 14*(4), 112–14.

Rust, T., Diessner, R., and Reade, L. (2009). Strengths Only or Strengths and Relative Weaknesses? A Preliminary Study. *Journal of Psychology, 143*(5), 465–76.

Scheier, M. F., Mathews, K. A., Owens, J. F., Magovern, G. J., Lefebvre, R. C., Abbott, R. A., and Carver, C. S. (1989). Dispositional Optimism and Recovery from Coronary Artery Bypass Surgery: The

Beneficial Effects on Positive Physical and Psychological Well-being. *Journal of Personality and Social Psychology, 57*(6), 1024–40.

Seligman, M. and Peterson, C. (2000). *Values in Action (VIA) Classification of Strengths*. Retrieved from http://www.positivepsychology.org/toxonomy/htm.

Seligman, M. (2008). Positive Health. *Applied Psychology: An International Review, 57*, 3–18.

Seligman, M., Park, N., Peterson, C. (2005). VIA Institute on Character: Research Abstracts. *Review of General Psychology, 9*(3), 203–13.

Seligman, M., Steen, T. A., Park, N., and Peterson, C. (2005). Positive Psychology Progress. Empirical Validation of Interventions. *American Psychologist, 60*(5), 410–21.

Sherbourne, C. D., Hays, R. D., Ordway, L., Dimatteo M. R., and Kravitz, R. L. (1992). Antecedents of a Adherence to Medical Recommendations: Results from the Medical Outcomes Study. *Journal of Behavioural Medicine, 15*(5), 447–68.

Singh, K., and Choubisa, R. (2010). Empirical Validations of Values in Action-inventory of Strength (VIS-IS) in Indian Context. *Psychological Studies, 55*(2), 151–58.

Swift, T. L., Ashcraft, R. E., Todd, W., Campbell, A. V., and Dieppe, P. A. (2002). Living Well Through Chronic Illness; the Relevance of Virtue Theory to Patients with Chronic Osteoarthritis. *Arthritis and Rheumatology: Arthritis Care and Research, 47*(5), 474–78.

Taylor, S. E., Kemeny, M. E., Reed, G. M., Bower, J. E., and Gruenewald, T. L. (2000). Psychological Resources, Positive Illusions, and Health. *American Psychologist, 55*(1), 99–109.

Tuck, I., McCain, N. L., and Elswick, R. K., Jr. (2001). Spirituality and Psychosocial Factors in Persons Living with HIV. *Journal of Advance Neuroscience, 33*(6), 776–83.

Van Eeden, C., Wissing, M. P., Dreyer, J., Park, N., and Peterson, C. (2008). Validation of the Values in Action Inventory of Strengths for Youth Among South African Learners. *Journal of Psychology in Africa, 18*(1), 145–56.

Wong, P. T. P. (2012). *The Human Quest for Meaning: Theories, Research, and Applications* (2nd Ed.). New York, NY: Routledge.

CHAPTER 5

Happiness and Quality of Life

Hardeep Lal Joshi

Kurukshetra University, Haryana

Abstract: Today's world is marked by some distinct characteristics, such as natural calamities, economic slowdown, selfishness, growing religious hatred, terrorism and growing aggression. In this world of negativity and sadness, the aspect which is most important is change in the basic attitude towards life, such as positivity and happiness. This change towards happiness will lead to improvement in the quality of life. The present chapter deals with happiness and how it is related to quality of life. In the beginning, the chapter deals with the similarity and differences between these aspects and some other aspects of positive psychology. Then the author focuses on the different factors determining these aspects. The present chapter discusses the various dimensions and factors affecting happiness and quality of life with some empirical studies. The main factors which are focused upon are emotional intelligence, social skills, self efficacy, social support and forgiveness. It is found that emotional intelligence has a significant positive relationship with happiness. Individuals with high level of self-esteem and efficacy are found to be high on happiness along with quality of life. A large number of studies conducted recently show that there is positive relationship between social skills and social support with happiness and quality of life. A new area of research, i.e., forgiveness is also related to happiness and quality of life. The present chapter provides an empirical record of the relationship between these two concepts. In the latter half, the chapter deals with one other dimension—the relationship between money and happiness. The chapter provides a critical review on this relationship. It is generally said that money brings happiness, but the World Happiness Index shows that the wealthiest people are not the happiest. In conclusion, the chapter evaluates how these concepts of happiness and quality of life are really thriving the individual as well as the whole society towards well-being which lead to a win–win situation not only for individuals, but society as a whole.

Happiness and Quality of Life

Throughout history, philosophers and religious leaders always showed their interests in knowing what makes man happy. How can a man be happy in adverse situations? They have suggested that diverse characteristics, such as love, wisdom and non-attachment are the cardinal elements of a fulfilled existence. Utilitarians like Jeremy Bentham (1789) argued that the presence of pleasure and the absence of pain are the defining characteristics of a good life. Thus, the Utilitarians were the intellectual forerunners of happiness researches, focusing on the emotional, mental and physical pleasures and pain that individuals experience. Although there are other desirable personal characteristics apart from a person's happiness the individual with abundant joy has one key ingredient of a good life.

Though happiness is very much related to the existence of human being, the empirical work on happiness actually started in 1925 by Flugel who studied mood states of the individual during emotional events (Flugel, 1925). So, Flugel was the forerunner as far as the approaches to measure happiness or subjective well-being are concerned. After World War II, many experts started using survey researches to measure happiness, life satisfaction and subjective well-being through objective measures. In literature, subjective well-being, happiness and life satisfaction are used synonymously. The Pollsters studied a large number of people who were often selected to produce representative samples of nations. One of the main works is conducted by George Gallup and his colleagues, and Hadley Cantril pioneered the use of large-scale surveys as an assessment technique (as cited by Diener, 2000). In their survey, people were given statements which they have to respond from 'very happy' to 'not so happy'. The composite score on all the statements shows the overall happiness. Then, experts in this area tried to formulate a national index for happiness (Diener, 2000).

Concept of Happiness

The word 'happiness' is used in various ways. In the widest sense, it is an all-inclusive term covering all that is good. It is often used interchangeably with terms like 'well-being' or 'quality of life' and denotes both individual and social welfare. The word, 'simply' means that there is one ultimate good and plays down differences in interests between individuals and society. The word, 'happiness' is now used in the more limited sense of subjective enjoyment of life.

Overall, happiness is the degree to which an individual judge the overall quality of his/her own life-as-a-whole favorably. The word, 'happiness' is not used to denote an optimal appreciation of life. It refers to a degree, like the concepts of 'length' or 'weight' which denotes more or less of something. The term, 'happiness' is used to describe the state of an individual only; it does not apply to collectivities. This term is used where somebody has made an overall judgement about the quality of his or her life. In this way, it is an intellectual activity. The evaluation of life aimed at is an overall judgement. It embodies all the criteria for appreciation, which figure in the mind. The word, 'happiness' cannot be used to characterize satisfaction with specific aspects of life, such as marriage or work. 'Happiness' refers to satisfaction with life-as-a-whole. It covers past, present and anticipated experiences. Evaluation of life always embodies appreciation; a conclusion as to whether one likes something or not.

Ouality of Life

Today the issues on quality of life are discussed widely in different scientific fields. Researchers in different branches of social sciences take the term 'quality of life' in different ways. In sociology, quality of life is understood in terms of a person's understanding of well-being taking into account his or her own needs. Economics looks upon it as a standard of living perceived by an individual. In medical sciences, it is lack of illness and more of health as an indicator towards better style of life.

There is no universally accepted definition of quality of life. Usually the definition provided by the World Health Organization (WHO) introduced in 1995 seems to be the most comprehensive one. It states that Quality of Life (QOL) is individuals' perception of their position in life in the context of the culture and value systems in which they live. QOL is related to goals, expectations, values and concerns of the individual incorporating his physical health, psychological state, level of independence, social relations, personal beliefs and their relationship. These are the salient features of the environment which refer to a subjective evaluation that is embedded in a cultural, social and environmental context [World Health Organization Quality of Life (WHOQOL) Group, 1995].

Theoretical Underpinning

Historically and from a theoretical point of view, happiness can be assessed from different angles. Experts are not unanimous whether there can be a single theory to evaluate happiness as a whole. Many thinkers have different points of view regarding their interpretation about happiness. Buddha, who was born in a royal family, left home in search of a more meaningful existence and ultimately found enlightenment, a sense of peace, and happiness. Philosophers like Aristotle believed that Eudaimonia (i.e., human flourishing associated with living a life of virtue), or happiness based on a lifelong pursuit of meaningful, developmental goals (i.e., 'doing what is worth doing'), was the key to the good life (Waterman, 1993). The founders of America reasoned that happiness was as important as our rights of life and liberty. These arguments and opinion of past and present philosophers and experts have a clear and remarkable impact on our present understanding and conceptualization of happiness. More recently, psychological theories along with genetic researches help us to understand happiness along with its correlates in a better way.

Many theories of happiness have been proposed since Aristotle's brilliant insights. These theories can be categorized into three groups, viz., (1) need and goal satisfaction theory; (2) process or activity theories; and (3) genetic and personality predisposition theories. The first constellation of theories centres around the idea that happiness is related to the reduction of tension (e.g., the elimination of pain and suffering and enhancement of biological and psychological needs leads to happiness). Psychologists like Freud and Maslow conceptualized it in different ways. Freud (1976) related happiness with immediate gratification of need that is the main function of Id which is governed by pleasure principle. Maslow (1970), main worker of humanistic-existential model, proposed happiness in context with needs which are presented in a hierarchy. Satisfaction with lower levels of need moves the individual up in the hierarchy. In support of this view, Omodei and Wearing (1990) found that the degree to which individuals' needs were met was positively associated with the degree of their life satisfaction.

The second line of theories is 'Goal Theories'. The goal theories state that more we move towards the goal or standard we have set for ourselves, happier and satisfied we become. Other researchers have extended this idea to incorporate the degree of discrepancy from other potential comparison standards. For example, Michalos (1985) postulated that happiness is inversely related to the degree of discrepancy from multiple standards. More the discrepancy between our real and expected goals, more problems we will face and less happy we will be. Higgins (1987) posited that broad discrepancies between *ideal* self and *ought* self lead to the experiences of negative emotions. One implication of tension reduction theories is that happiness occurs after needs are met and goals are fulfilled. In other words, happiness is a desired end towards which all activities are directed. These theories can be compared with models of happiness in which engagement in an activity itself provides happiness. Most notably, Csikszentmihalyi (1975) suggested that people are happiest when they are engaged in interesting activities that match their level of skills. He explained that whatever the activities we are doing, if it matches with our level, it provides happiness. He termed it skill 'flow' and argued that more this 'flow' is experienced by the people, happier they are. Similarly, Cantor and her colleagues (Cantor and Blanton, 1996; Harlow and Cantor, 1996) found retired elders' participation in social activities as potent predictors of happiness and life satisfaction. They postulated that it is not the participation, but active participation on the part of the individuals in life tasks which makes them happy. Sheldon, Ryan and Reis (1996) found that it is the intrinsic reasons for engaging in activities which make people happy. As far as the indicators of well-being are concerned, goal theorists like Emmons (1986) and Little (1989) agreed that having important goals and pursuing them with determination make people feel psychologically healthy and equipped. Happiness is not a transient phenomenon rather it is consistent and long lasting. This consistency is found in various studies. Magnus and Diener (1991) found high correlation between life satisfaction measures over a period of four years. Similarly, Costa and McCrae (1988) reported stability in happiness especially in affective component for a period of six years.

These results clearly pointed that life events in one's life influence one's happiness, but along with these, environmental, biological and genetic factors too play an important role. Diener, Sandvik, Sedlitz and Diener (1993) conducted a study on people who experienced some change in their income level in the last ten years. They found that there is stability as far as happiness and subjective well-being are concerned. These results have led some theorists to suggest that although life events can influence subjective well-being, people eventually adapt to these changes and return to biologically determined 'set points' or 'adaptation levels' (Headey and Wearing, 1992).

One reason for the stability and consistency of happiness is that there is a substantial genetic component to it; to some degree people are born prone to be happy or unhappy. In a study conducted by Tellegen *et al.* (1988) monozygotic twins, who were reared apart, were examined and compared with dizygotic twins who were reared apart as well, and raised together with monozygotic and dizygotic twins, after comparing the similarities of the various types of twins. Tellegen *et al.* (1988) found that genetic predispositions among twins predicted 40 per cent of the variability in positive emotions and 55 per cent of the variability in negative emotions. It shows that genes are the main determinants of emotional responses as compared to environmental factors.

When personality theorists tried to find out the most consistent traits that are linked to happiness and subjective well-being, they found two extraversion and neuroticism (Diener and Lucas, 1999). In a meta-analysis, Lucas and Fujita (2000) found that extraversion is positively and moderately correlated with pleasant affect whereas neuroticism and negative affect are positively correlated (Fujita, 1991). Another model of personality, i.e., five factor model relates traits like agreeableness, conscientiousness and openness to experiences positively with happiness and subjective well-being. But the relationship is less consistent and smaller as compared to broad dimensions like extraversion and neuroticism (Watson and Clark, 1992). People's cognition about the world and humans also had some relationship with happiness. Certain cognitive dispositions like hope (Snyder *et al.*, 1991), dispositional optimism (Scheier and Carver, 1993) and expectancy for control (Grob, Stetsenko, Sabatier, Botcheva and Macek, 1999) appear to influence happiness. So, what matters in happiness is not what we are, but also what we think about life and about others.

Personality Traits and Happiness

From ancient times, philosophers and other experts in different areas were of the view that there exist some differences among happy and unhappy people. Studies in the field of personality show that personality profiles of happy as well as unhappy people are different. Happy people have extravert tendencies like optimism, high self-esteem and better internal locus of control. On the other hand, unhappy people are high on neuroticism; have poor self-esteem and high external locus of control.

The associations between personality traits and happiness are not universal across all cultures. There is considerable evidence from longitudinal studies of the link between temperamental and personality traits. It is found that children with high activity levels and positive affect become extraverted, and so are more likely to be happy. Children who are highly irritable and fearful show high levels of neuroticism in later life and so are more likely to show negative affectivity (Rothbart and Ahadi, 1994). Research shows that secure attachment patterns which are developed by children due to their interactions with significant people in their life determine happiness in their later life.

Though studies show that optimism, self-esteem and locus of control are some of the personality traits which correlate with happiness, yet the relative contributions of genetic and environmental factors to the development of these traits are less clear. However, there are good reasons to believe that secure attachment is an important precursor of these types of personal strengths (Thompson, 1999). Landmark studies in the area of emotional intelligence reveal that high emotional intelligence would lead to greater feelings of emotional well-being (Goleman, 1995; Saarni, 1999; Salovey and Mayer, 1990; Salovey, Mayer, Goldman, Turvey and Palfai, 1995). Those who are able to understand and regulate their emotions should be able to generally maintain a better outlook on life and experience better emotional health. Some empirical studies show that higher levels of emotional intelligence is associated with lower levels of depression (Martinez–Pons, 1997; Schutte *et al.*, 1998), higher hope and optimism (Schutte *et al.*, 1998) and greater life satisfaction (Austin, Saklofske and Egan, 2005; Ciarrochi, Chan and Caputi, 2000; Palmer, Donaldson and Stough, 2002).

Culture and Happiness

Several factors related to culture and socio-political environment have also been found to play important role in determining happiness (Triandis, 2000). In cross-cultural studies, associations have consistently been found between subjective well-being and living in a stable democracy. Democratic set up in society has been found to be positively associated with happiness. People in those cultures are found to be happier where there is importance given to welfare of the people, less bias on the part of the administration and satisfactory relationship between citizens and bureaucracy.

Diener (1994) found marked differences among nations on happiness when their income differences are controlled. It is found that the Swiss ranked themselves higher on self-report measure of life satisfaction than their German, French and Italian counterparts (Inglehart, 1990).

Environment and Happiness

It is still not clear whether pleasant physical environment in which individuals live has any impact or relationship with happiness. This association is quite complex. Studies conducted at different adverse physical environments show that the people in these areas are found to be happy as compared to people who live in luxurious environment.

Wealth

It is an observation by general population that wealthy people are happier than less wealthy people. Diener (2000) found that people living in economically poor countries set their happiness at low levels. Across nations, there is high correlation (.60) between happiness and wealth. Happiness is low in poor countries because people in those countries perceive that government has not provided them with the luxuries they require.

In economically advanced countries, people who value earning money more than other goals are less satisfied with their standard of living and their lives (Myers, 2000; Sirgy, 1998). This may be because accumulation of money may not be conducive in meeting the social and psychological needs that enhance happiness. This negative correlation between happiness and pursuit of money has not been studied in economically disadvantaged countries.

Geographical Location and Other Factors

As far as the geographical location is concerned, positive feelings are associated with natural rather than artificial environments. People have strong liking for those geographical locations where there is vegetation, water and panoramic views (Ulrich, Dimberg and Driver, 1991). Buss (2000) related this association to evolutionary factors. We are more close to nature that is why we prefer these locations. Cunningham (1979) reported that positive moods are associated with good weather. Environment induces positive moods where natural music flows. Surveys and many mood induction experiments show that music induces some effect on positive mood states and reduces aggression (Argyle, 2001).

Physical State and Happiness

Several studies have been conducted to find out the relationship between physical state and happiness. Areas studied are put under different headings:

Health

Health is found to be associated with happiness. Most of the normal people adapt to their health related problems with the passage of time. They later develop self-perception of their health which shows consistency with their happiness level. Now, a new line of research has emerged which examines the role of happiness in our health as well as its impact on our immune system. Studies showed that immune system of happy people works more effectively as compared to unhappy people (Kamen–Siegel, Rodin, Seligeman and Dwyer, 1991;

Segertrom, Taylor, Kemeny and Fahey, 1998; Stone *et al.*, 1994). Those who show positive emotions, their tolerance of pain increases (Weisenberg, Raz and Hener, 1998).

Exercise

It is concluded that short-term exercise induces positive mood states and long-term regular exercise leads to happiness (Argyle, 2001; Sarafino, 2002). The short-term effects of exercise release high level of endorphins and other substances in the brain. Regular physical exercise reduces depression and anxiety, enhances speed and accuracy in our performance, improves our self-esteem and our physical fitness, thereby leading to better cardiovascular functioning.

Productivity and Happiness

Productivity and happiness are inter-related, as higher level of productivity leads to higher level of happiness. This type of relationship is evident when the work is according to our interest and provides us satisfaction. So, employment status, job satisfaction and skill usage with goal-directed activities are associated with happiness. The relationship of happiness with various factors has been studied as follows:

Work

Employment status of the individual is related to happiness. Argyle (2001) and Oswald (1997) found that employed people are happier than unemployed ones, and people engaged in professional and skilled jobs are happier as compared to their unskilled counterparts. Happiness and job satisfaction are positively correlated (Diener, Suh, Lucas and Smith, 1999). The reason for this is that work provides people an opportunity to boost their self-worth, for enhancing their skills, developing a support-based system and getting pleasure out of it. As far as the choice regarding job is concerned, people opt for those jobs which are satisfying, provide good physical and psychological environment and have distinctive characteristics (Warr, 1999). The satisfied and happy people prefer those jobs which suit their talent, skills and preferences.

In a meta-analysis, Tait, Padgett and Baldwin (1989) found moderate positive correlation (.44) between job satisfaction and life satisfaction. One new finding of the study is that there is stronger relationship between job and life satisfaction among females too which was not found earlier. Tait *et al.* (1989) explained that it is because of their changing societal role and extended choice. Studies revealed that intrinsic rewards and social benefits from jobs are significant predictors of satisfaction. Later research shows that personality factors are the main underlying factors whether it comes to general satisfaction from life or satisfaction at job level.

Education

Education level is positively correlated with happiness. Diener *et al.* (1999) found that the relation is strong enough among people in low-income groups in developed nations and general population among third world countries. This may be because higher education in underdeveloped countries provides greater benefits as compared to low education. Witter, Okun, Stock and Haring (1984) in their meta-analysis found significant, but low effect size (.13) between education and happiness. This effect size is similar to other areas, such as life satisfaction (.15), morale (.15) and quality of life (.12). Education is positively related to happiness in low-income groups and poor countries because they associate education as the provider of better facilities and leisure. Interestingly, developed countries like America observed poor relationship between education and happiness. Effect of education became weaker over time in America. Most of the studies focusing on exploring the relationship between education and happiness have been limited to education with occupational status and income (Witter, Okun, Stock and Haring, 1984). Hence, more studies are required to examine this relationship from other perspectives as well.

Goal Attainment

Attaining goals which people have set for themselves are one of the factors related to happiness. People report greater happiness when they achieve highly valued goals than when they achieve less valued goals (Diener *et al.*, 1999). These goals may be determined by many factors including personality traits. In keeping with this, extrovert people experience more happiness when they do exciting things and introverts experience more happiness when they do things that bring contentment to them.

Negative Social Comparison and Happiness

Our level of happiness is determined and influenced by how we rate ourselves and our current circumstances in comparison with others (Wood, 1996). We compare ourselves with other people in terms of health, wealth, social status, academic and athletic achievement, personal attractiveness and that of our partners and children, and so forth. When we compare ourselves with people who are very high as compared to us, the greater the gap, greater the gloom we suffer.

Whether Rich are Happier than Poor

'Whether money brings happiness' is an obvious question for debate. People have a general feeling that where there is enough money there is happiness. A large number of studies have been carried out by psychologists to find out the relationship between these two variables. Overall, a small, but significant correlation of .12 between income and happiness or subjective well-being was found in a nationally representative sample in the United States

(Diener *et al.*, 1993). The effect of income on happiness is small even in the case of high earners. Individuals with very high income were compared with low to moderate income group counterparts and the geographical effect was partialled out.

Results showed that the rich, on average, appeared somewhat happier than the mean of national samples, but there was considerable overlap between wealthy and not wealthy groups. Generally, wealthy people are happier than poor people; however, the effects are small. It was studied that amongst individuals who won large amounts of money, only 39 per cent described themselves very happy as compared to 19 per cent of the controls. Unexpectedly, some negative effect also surfaced as the winners quit their jobs and then experienced lost relationships and decreased feelings of accomplishment, as well as some tension from family and friends who expected financial assistance. Diener *et al.*, (1993) found no differences in pleasure level between groups undergone at least one-half a standard deviation in income over a ten-year period. Income change at the national level, as with the personal level, shows little effect upon subjective well-being. Though strong economic growth has been experienced in France, Japan and the United States from 1946–90, there was change in mean reports of happiness (or subjective well-being) (Diener and Suh, 1997). Disposable income rose dramatically (even after controlling for inflation and taxes), but happiness and well-being levels remained stable. Oswald (1997) studied nine European nations during a time of rapid economic growth and also found virtually no increase in happiness. Contrarily, Gross National Product (GNP) per capita correlated approximately .50 with life satisfaction across thirty-nine nations. These results were supported by a sample of fifty-five nations using both GNP and purchasing power as indicators of national financial status (Diener and Diener, 1995). This relationship between wealth and subjective well-being is not because of wealth alone but may be due to other associated factors like democratic and egalitarian treatment of the individual living in developed and wealthy nations.

Thus, findings do not support a strong causal direction from income to happiness. Wealthy people are only somewhat happier than poor people in rich nations, though, wealthy nations appear much happier than poor nations, and changes in income may sometimes have negative effect. Studying expectations and material desires would be beneficial in explaining these patterns. For example, wealth affects happiness more so in countries where basic needs of food, shelter and health care are of concern, but has a much lesser effect within countries where basic needs are met. People working for long hours may get more money, but their level of satisfaction and happiness is somewhat compromised. The sacrifice of important values, such as self-growth, leisure time and significant relationships to attain more wealth may interfere with happiness rather than enhance it. In question is not the pursuit of wealth and economic growth of nations and individuals, rather, how increases in wealth may be used and at what costs.

Conclusion

It can be concluded that happiness is, nowadays, one of the most important priorities of human life. It is a relative term and not an absolute one. It all depends upon one's judgement of life. Happiness is an umbrella term which includes quality of life. Though happiness is discussed from the very beginning of human existence, yet no theory in the past or present is able to conceptualize happiness in total. There are various factors which are found to be related to happiness such as personality factors, work environment, productivity, physical state, education, and so forth. But more research studies are required to have a holistic view of happiness. The debates point to the fact that money and happiness have some relationship, but this relationship is not well established. Natives of some of the poor nations are found to be happier as compared to their counterparts from the wealthy nations. So, it can be concluded that wealth is not the only factor which brings happiness; one's evaluation about the situation is far more important.

References

Argyle, M. (2001). *The Psychology of Happiness*. London: Routledge.

Austin, E. J., Saklofske, D. H., and Egan, V. (2005). Personality, Well-being, and Health Correlates of Trait Emotional Intelligence. *Personality and Individual Differences*, *38*(3), 547–58.

Bentham, J. (1789). *Introduction to the Principles and Morals of Legislation*. London: University of London Athlone Press.

Buss, D. (2000). The Evolution of Happiness. *American Psychologist, 55*(1), 15–23.

Cantor, N., and Blanton, H. (1996). Effortful Pursuit of Personal Goals in Daily Life. In J. A. Bargh and P. M. Gollwitzer (Eds.). *The Psychology of Action: Linking Cognition and Motivation to Behaviour* (pp. 338–359). New York: Guilford.

Ciarrochi, J. V., Chan, A. Y., and Caputi, P. (2000). A Critical Evaluation of the Emotional Intelligence Construct. *Personality and Individual Differences*, *28*(3), 539–61.

Costa P. T., McCrae, R. R., and Zonderman, A. (1987) Environmental and Dispositional Influences on Well-being: Longitudinal Follow-up of an American National Sample. *British Journal of Psychology, 78*(3), 299–306.

Costa, P. T., and McCrae, R. R. (1988). Personality in Adulthood: A Six Year Longitudinal Study of Self-reports and Spouse Rating on the NEO Personality Inventory. *Journal of Personality and Social Psychology*, *54*(5), 853–63.

Csikszentamihalyi, M. (1975). *Beyond Boredom and Anxiety*. San Francisco: Jossey-Bass.

Csikszentmihaiyi, M. (2000). Happiness, Flow, and Economic Equality. *American Psychologist, 55*(10), 1163–64.

Cunningham, M. (1979). Weather, Mood and Helping Behaviour: Quasi-experiments with the Sunshine Samaritans. *Journal of Personality and Social Psychology*, *37*(11), 1947–56.

Diener, E., and Lucas, R. (1999). Personality and Subjective Well-being. In E. Kahneman, E. Diener and N. Schwartz (Eds.), *Well-being: The Foundations of Hedonic Psychology* (pp. 213–29). New York: Russell Sage Foundation.

Diener, E. (1994). Assessing Subjective Well-being: Progress and Opportunities. *Social Indicators Research, 3*(2), 103–57.

Diener, E. (2000). Subjective Well-being: The Science of Happiness and a Proposal for a National Index. *American Psychologist*, *55*(1), 34–43.

Diener, E., and Suh, E. (1997). Measuring Quality of Life: Economic, Social, and Subjective Indicators. *Social Indicators Research, 40*(1–2), 189–216.

Diener, E., Suh, E. M., Lucas, R. E., and Smith, H. L. (1999). Subjective Well-being: Three Decades of Progress. *Psychological Bulletin, 125*(2), 276–302.

Diener, E.. and Diener, M. (1995). Cross-cultural Correlates of Life Satisfaction and Self-esteem. *Journal of Personality and Social Psychology*, *68*(4), 653–63.

Diener, E., Sandvik, E., Sedlitz, L., and Diener, M. (1993). The Relationship between Income and SWB: Relative or Absolute *Social Indicators Research*, *28*(3), 195–223.

Emmons, R. A. (1986). Personal Strivings: An Approach to Personality and SWB. *Journal of Personality and Social Psychology*, *51*(5), 1058–68.

Flugel, J. C. (1925). A Quantitative Study of Feeling and Emotion in Everyday Life. *British Journal of Psychology*, *15*(4), 318–55.

Freud, S. (1976). *New Introductory Lectures on Psychoanalysis*. In J. Strachey (Ed. and Trans). The Complete Psychological Works (Vol. 16, pp. 9–16). New York: Norton (Original Work Published 1933).

Fujita, F. (1991). *An Investigation of the Relation Between Extraversioin, Neuroticisism, Positive Affect and Negative Affect*. Unpublished Master's Thesis. University of Illinois, Urbana-Champaign.

Goleman, D. (1995). *Emotional Intelligence*. New York: Bantam.

Grob, A., Stetsenko, A., Sabatier, C., Botcheva, L., and Macek, P. (1999). A Cross-national Model of SWB in Adolescence. In F. D. Alsaker and A. Flammer (Eds.) *The Adolescents in the 1990s. Research Monographs in Adolescence* (pp. 115–30). Mahwah, NJ. Erlbaum.

Harlow, R. E., and Cantor. N. (1996). Still Participating After all These Years: A Study of Life Task Participation in Later Life. *Journal of Personality and Social Psychology*, *71*(6), 1235–49.

Headey. B., and Wearing, A. (1992). *Understanding Happiness: A Theory of SWB*. Melbourne: Longman Cheshire.

Higgins, E. T. (1987). Self-discrepancy: A Theory Relation Self and Affect. *Psychological Review*, *94*(3), 319–40.

Inglehart, R. (1990). *Culture Shift in Advanced Industrial Society*. Princeton, NJ: Princeton University Press.

Kamen-Siegel, L., Rodin, J., Seligeman, M., and Dwyer, C. (1991). Explanatory Style and Cell Mediated Immunity. *Health Psychology*, *10*(4), 229–35.

Little, B. R. (1989). Personal Projects Analysis: Trivial Pursuits, Magnificent Obsessions and the Search for Coherence. In D. Buss and N. Cantor (Eds). *Personality Psychology: Recent Trends and Emerging Directions* (pp. 15–31). New York: Springer-Verlag.

Lucas, R. E., and Fujita, F. (2000). Factors Influencing the Relation Between Extraversion and Pleasant Affect. *Journal of Personality and Social Psychology*, *79*(6), 1039–56.

Magnus, K., and Diener, E. (1991). *A Longitudinal Analysis of Personality, Life Events and SWB*. Paper Presented at the 63rd Annual Meeting of the Midwestern Psychological Association, Chicago.

Martinez-Pons, M. (1997). The Relation of Emotional Intelligence with Selected Areas of Personal Functioning. *Imagination, Cognition, and Personality*, *17*(1), 3–13.

Maslow, A. H. (1970). *Motivation and Personality*. New York: Harper and Row.

Michalos, A. C. (1985). Multiple Discrepancies Theory (MDT). *Social Indicators Research*, *16*(4), 347–413.

Myers, D. (2000). The Funds, Friends and Faith of Happy People. *American Psychologist*, *55*(1), 56–67.

Omodei, M. M., and Wearing A. J. (1990). Need Satisfaction and Involvement in Personal Projects toward an Integrative Model of Subjective Well-being. *Journal of Personality and Social Psychology*, *59*(4), 762–69.

Oswald, A. J. (1997). Happiness and Economic Performance. *Economic Journal, 107*(445), 1815–31.

Palmer, B., Donaldson, C., and Stough, C. (2002). Emotional Intelligence and Life Satisfaction. *Personality and Individual Differences*, *33*(7), 1091–1100.

Phillips, D. (2006). *Quality of Life: Concept, Policy and Practice.* New York: Routledge.

Rothbart, M., and Ahadi, A. (1994). Temperament and the Development of Personality. *Journal of Abnormal Psychology, 103*(1), 55–66.

Saarni, C. (1999). *The Development of Emotional Competence.* New York: Guilford Press.

Salovey, P., and Mayer, J. D. (1990). Emotional Intelligence. *Imagination, Cognition, and Personality*, 9(3), 185–11.

Salovey, P., Mayer, J. D., Goldman, S. L., Turvey, C., and Palfai, T. P. (1995). Emotional Attention, Clarity and Repair: Exploring Emotional Intelligence Using the Trait Meta-Mood Scale. In J. W. Pennnebaker (Ed.), *Emotion, Disclosure, Health* (pp. 125–54). Washington, DC: American Psychological Association.

Sarafino, E. (2002). *Health Psychology* (4th ed). New York: Wiley.

Scheier. M. F., and Carver, C. S. (1993). On the Power of Positive Thinking. The Benefits of Being Optimistic. *Current Directions in Psychological Science*, *2*(1), 26–30.

Schutte, N. S., Malouff, J. M., Hall, L. E., Haggerty, D., Cooper, J. T., Golden, C., and Dornheim, L. (1998). Development and Validation of a Measure of Emotional Intelligence. *Personality and Individual Differences*, *25*(2), 167–77.

Segertrom, S., Taylor, S., Kemeny, M., and Fahey, J. (1998). Optimism is Associated with Mood, Coping, and Immune Change in Response to Stress. *Journal of Personality and Social Psychology*, *74*(6), 1646–55.

Sheldon, K. M., Ryan. R., and Reis, H. T. (1996). What Makes for a Good Day? Competence and Autonomy in the Day and in the Person. *Personality and Social Psychology Bulletin*, *22*(12), 1270–79.

Sirgy, M. (1998). Materialism and Quality of Life. *Social Indicators Research*, *43*(3), 227–60.

Snyder, C. R., Harris, C., Anderson, J. R., Holleran, S. A., Irving, L. M., Sigmon, S. T.... Harney, P. (1991). This Will and the Ways: Development and Validation of an Individual Differences Measures of Hope. *Journal of Personality and Social Psychology*, *60*(4), 570–85.

Stone, A. A., Neale, J., Cox, D., Napoli, A., Valdimarsdottir, H. B., and Kennedy-Moore, E. (1994). Daily Events are Associated with Secretory Immune Responses to an Oral Antigen in Men. *Health Psychology, 13*(5), 440–46.

Tait, M., Padgett, M. Y., and Baldwin, T. T. (1989). Job and Life Satisfaction: A Review and Reevaluation of the Strength of the Relationship and Gender Effects. *Journal of Applied Psychology, 74*(3), 502–07.

Tellegen, A., Lukken, D. T., Bouchard, T. J., Wileox, K. J., Segal, N. L., and Rich, S. (1988). Personality Similarity in Twins Reared Apart and Together. *Journal of Personality and Social Psychology*, *54*(6), 1031–39.

Thompson, R. (1999). Early Attachment and Later Development. In J. Cassidy and P. Shaver (Eds). *Handbook of Attachment* (pp. 265–286). New York: Guilford.

Triandis, H. (2000). Cultural Syndromes and Subjective Well-being. In E. Diener and E. Suh (Eds). *Culture and Subjective Well-being* (pp. 13–36). Cambridge, MA: MIT Press.

Ulrich, R., Dimberg, U., and Driver, B. (1991). Psycho-physiological Indicators of Leisure Benefits. In B. Driver, P. Brown and G. Peterson (Eds)., *Benefits of Leisure* (pp. 73–89). State College, PA: Venture Publishing.

Warr, P. (1999). Well-being and the Workplace. In E. Kahneman, E. Diener and N. Schwartz (Eds.), Well-being: *The Foundations of Hedonic Psychology* (pp. 393–412). New York: Russell Sage Foundation.

Waterman, A. S. (1993). Two Conceptions of Happiness: Contrasts of Personal Expressiveness (Eudaemonia) and Hedonic Enjoyment. *Journal of Personality and Social Psychology*, *64*(4), 678–91.

Watson, D., and Clark, L. A. (1992). On Traits and Temperament: General and Specific Factors of Emotional Experience and their Relation to the Five Factor Model. *Journal of Personality*, *60*(2), 441–76.

Weisenberg, M., Raz, T., and Hener, T. (1998). The Influence of Film Induced Mood on Pain Perception. *Pain*, *76*(3), 365–75.

WHOQOL Group (1995). The World Health Organization Quality of Life Assessment (WHOQOL): Position Paper from the World Health Organization. *Social Science and Medicine, 41*(10), 1403–09.

Witter, R. A., Okun, M. A., Stock, W. A., and Haring, M. J. (1984). Education and Subjective Well-being: A Meta-analysis. *Educational Evaluation and Policy Analysis*, *6*(2), 165–73.

Wood, J. (1996). What is Social Comparison and How Should We Study It? *Personality and Social Psychology Bulletin*, *22*(5), 520–37.

CHAPTER 6

An Analytical Model of Happiness

Akbar Husain and Shahin Zehra
Department of Psychology, Aligarh Muslim University, Uttar Pradesh

Abstract: In this chapter, the authors have attempted to provide a clear explanation of the concept 'happiness'. Not only the definition is given to explain just what happiness is, several studies on happiness and its psycho-social correlates are quoted to demonstrate its originality. This includes the most important common points of reference when understanding the relationship of happiness with psycho-spiritual concepts that fits easily under the operational definition of happiness. The authors hold that there is a need to look at life and happiness in a practical way. Happiness can be defined as composite of health, attributional style, pleasure, positive emotions, interpersonal relationships, nature, extraversion, spirituality and self-esteem. Considering *happiness* as an acronym, this model can be used to develop a broad theory that explains the concept of happiness.

An Analytical Model of Happiness

Happiness is a state of mind. When the mind is tranquil, you are happy. We are in a constant state of unrest that arises from the feelings of insecurity and inadequacy. The mind gets confused when it is full of unfulfilled desires. We have unrealistic desires and aspirations without realizing their adverse effects on our health. All religions warn against excessive desire of self that leads to destructive behaviour. We are in the mindless pursuit of fulfilling desires and, hence, feel unhappy. That is why we never feel happy.

William James suggested, 'how to keep, how to gain, how to recover happiness is, … for most men, at all times, the secret motive for all they do' (1902/1958, p. 76). Most people agree that being happy is the ultimate goal toward which they strive. Many people strive for material possessions throughout their life. But, there is no guarantee that even after getting those possessions or attaining those positions, they would be happy. For instance, Diener and Oishi (2004), reported that being happy was rated to be more important than having good health, a high income, or high levels of attractiveness; and it was rated as being more important than experiencing love or meaning in life. Thus, happiness is seen as an ultimate goal that guides individual choices and, which can be achieved if the external circumstances in a person's life match with his or her desires (Ryff and Keyes, 1995).

White (2006) paid attention to the factors that influence happiness–virtues, merits, values and 'what's desirable and worthwhile'. According to Martin (2005) those who have recognized these factors lead happier lives. Happiness is found in, and sometimes identified with, a life of fulfilment and harmony both within the individual and in that individual's relations with others (Macquarrie and Childress, 2001). People's behaviour is often guided by their beliefs about the types of things that will make them happy (Gilbert, 2006). People may choose a high paying job, an expensive house, or a short commute over other alternatives because they believe that these life circumstances will help them improve or maintain their happiness.

Happiness from the Viewpoint of Buddhism

Wallace and Shapiro (2006) observed that the Buddhist tradition has focused for over 2500 years on cultivating exceptional states of mental well-being as well as identifying and treating psychological problems. Recent research in neuroscience is the beginning to show support for Buddha's theory of happiness. According to Buddhism, the cause of happiness is virtuous *karma,* the law of causality and delusion. For instance, Davidson, Kabat-Zinn, Schumacher, Rosenkranz, Muller and Santorelli (2003) found that novice meditation practice was linked with significant greater activity in the left prefrontal cortex, an area of brain associated with positive emotion. They further supported it while examining the effects of mindfulness meditation on brain activity as well as psychological and immunological functioning. According to Buddhism, mental suffering is largely due to imbalances of the mind. Sayings of Buddhists (*Dhammpada*) are important in understanding the nature of happiness described under the headings: (1) The twin-verses; (2) Awareness; (3) Joy; and (4) Pleasure.

Happiness from the Viewpoint of Christianity

From the vantage point of view of Christianity, the following are among the factors that can cause human happiness: fearing God, trusting in the Lord, wisdom and understanding, righteousness, endurance, having mercy on the poor, keeping the law, being hopeful in the Lord, mourning, being merciful, purifying one's heart and being a peacemaker.

Happiness from the Viewpoint of Hinduism

The holy Gita offers us some practical tools to connect with our deepest intangible essence and we must learn to participate in the battle of life with right knowledge. The Gita teaches us how to get out of unhappiness or dissatisfactions of life by cultivating philosophy of life; identifying with the inner core of self-sufficiency; striving for excellence following the motto, work is worship; building up an internal integrated reference point to face contrary impulse and emotions; and pursuing ethico-moral rectitude.

The Hindu values enable us to cultivate happiness spiritually. They include firm faith in the following principles: (1) our divine heritage; (2) unity of existence; (3) freedom of conscience in religion; (4) sacredness of life; (5) sanctity of family life; (6) respect for parents and elders and reverence for the spiritual teacher; (7) the motherhood of god; (8) moderate pursuit of material welfare; (9) steadfast pursuit of 'self-integration and self-realization'; and, (10) renunciation and service for the good of the common people.

Happiness from the Viewpoint of Islam

The Qur'an says the following are among the factors that can be linked to human happiness: believing in God and Last Day, working righteousness, following God's guidance, spending one's wealth in the way of God, performing prayer and paying the alms, being God fearing, being among God's friends, doing good, not devouring usury, being patient and being steadfast.

Definitions of Happiness

A variety of biological, psychological, religious and philosophical approaches have made an attempt to define happiness and identify its sources. Philosophers and religious thinkers often define happiness in terms of living good life, or flourishing, rather than simply as an emotion. A number of personality traits are made use of to characterize important aspects of behaviour and experience. Particular personality trait profiles have been found to be associated with happiness which means happy and unhappy people have distinctive personality profiles. In Western cultures happy people are extraverted, optimistic, have high self-esteem and an internal locus of control. In contrast unhappy people tend to have high levels of neuroticism. The association between personality traits and happiness are not universal across all cultures.

Happiness is often defined empirically, i.e., via people's direct ratings of their happiness, long-term balance of positive and negative affect, or life-satisfaction (Diener, 1984; Lyubomirsky, 2001). Notably, these terms are not readily separable (Lyubomirsky and Lepper, 1999; Stones and Kozma, 1985), and thus we use them interchangeably. Happiness essentially relies on an individual's own perspective. Indeed, it would make little sense to pronounce a particular person as happy unless he thought so himself. Veenhoven (1989) defined happiness as the 'overall appreciation of one's life, as a whole' (p. 5). The most widely accepted definition of happiness is that of Diener and his colleagues, who prefer to use the label subjective well-being. They define happiness as a combination of life satisfaction (a cognitive judgement) and the balance of the frequency of positive and negative affect (i.e., hedonic tone) (Diener, Sandvik and Pavot, 1991; Larsen, Diener and Emmons, 1985). However, researchers widely differ in defining what happiness is. For example, some have focused on the individual's own mental or cognitive framework and perception.

The primary focus of this chapter is on a person's psychological characteristic associated with happiness during his or her life. Happiness may be defined in this way because the authors wish to identify the psychological correlates of happiness. Happiness involves healthy personality variables which are already researched. Happiness can be conceptualized as an outcome that reflects the conditions in a person's life. The authors suggest that happiness may be defined in terms of certain psychological factors, viz., health, attributional style, positive thinking, positive emotions, interpersonal relationships, nature, extraversion, social support and self-esteem. If we conceptualize happiness in this way then it can be assumed that everything will go well. Operationally, one might define a person's happiness level in terms of these psychological characteristics.

Health and Happiness

The theoretical literature suggests that happiness contributes to health and vice versa. Stones and Kozma (1980) suggested that happiness is a higher order construct that exerts an influence on lower order attitudinal and behavioural variables such as perceived health. Argyle (1987, 1997) clearly stated that health is closely linked with happiness. Health is widely regarded as one of the causes of happiness. Argyle argued that happiness probably causes health. Furthermore, Veenhoven (1989) proposed that happiness tends to foster health. He suggested that even when happiness does not promote health, it does reduce health complaints. Seligman (2002) proposed that subjective perception of health rather than objective good health is related to happiness. He suggested that we can appraise our subjective health positively even when we are quite sick. Only severe and long lasting illness brings unhappiness in its wake.

Health and happiness are important and possibly related pursuits of mankind (Graves, 2000). Health may be a determinant of happiness (Kozma and Stones, 1983). On the other hand, a feeling of happiness may enhance health. There is also evidence that health and subjective well-being may mutually influence each other, as good health tends to be associated with greater happiness (Okun, Stock, Haring and Witter, 1984). Happier people tend to have healthier lifestyles (Watson, 1988). Self-reported happiness is strongly related to health. Various studies have found that happiness is more strongly related to people's mental health than to their objective physical health. However, strong associations are found between happiness and people's perception of their own health, i.e., their so-called subjective health. Long-term studies indicate that the influence of subjective health on happiness becomes even stronger in old age.

Diener's (2008) study suggests that probing into one's happiness is one of the most important things a doctor can do to predict one's health and longevity. Majority of people overlook one's emotions as a vital component of one's health, while over focusing on diet and exercise. According to Diener and Biswas–Diener (2008), people who are happy become less sick than people who are unhappy. Diener (2008) suggests that there are three types of health: morbidity, survival and longevity. Evidence suggests that all three types of health can be improved through happiness. Morbidity, simply put, is succumbing to serious illness such as flu or cancer (Diener, 2008). In a 30-year longitudinal study, participants who were high in positive emotions were found to have lower rates of many health problems. Some of these illnesses/problems include lower death rates from heart disease, suicide, accidents, homicides, mental illnesses, drug dependency and liver disease related to alcoholism. Results also showed that depressed participants were more likely to have heart attacks and recurrences of heart attacks when compared to happy people. Survival is the term used for what happens to a person after he/she has already developed or contracted a serious illness (Diener, 2008). Happiness has been revealed to increase health, with survival. Evidence suggests that survival type of health may actually be sometimes detrimental to happiness. Danner, Snowdon and Friesen (2001) have made an attempt to determine a link between an individual's happiness and the individual's longevity. They recruited 180 Catholic nuns from a nearby convent as the participants of this study. Results showed that nuns who were considered happy or positive in their manner and language on average lived 10 years longer than the nuns who were considered unhappy or negative in their manner and language. A follow-up study was conducted by Pressman and Cohen (2012) who examined 96 famous psychologists. Results showed that the positive or happy psychologists lived, on average, six years longer. The psychologists who were considered negative or unhappy lived, on average, five years less.

Attributional Style and Happiness

Depressed people blame themselves for bad things that happen. However, this is not so much a cause of depression as an effect of it. Happy people attribute good events to themselves, not bad ones, but we don't know the direction of causation yet (Argyle, Martin and Crossland, 1989). Ryan and Deci (2001) pointed out that people high in happiness or subjective well-being tend to have attributional styles that are more self enhancing and more enabling than those low in subjective well-being.

Positive Thinking and Happiness

Happiness is characterized by positive affective states such as optimism, positive thinking and the perception of personal well-being. Happy people tend to think about and remember positive events in their lives. Unhappy people tend to dwell on negative happenings, and ruminate about their problems and distress (Lyubomirsky, 2001; Lyubomirsky and Ross, 1999). To attain happiness, we need to make a conscious effort to focus on our positive thoughts. Positive thinking is a stepping-stone to experience the divine vibe in meditation. Positive thinking helps in discovering the spiritual path and happiness. It helps to bring fresh and creative approaches to our practice. Our thoughts are the product of our values. Positive thinking has an impact on our personality. Human behaviour is unpredictable and dangerous if you do not think from the heart. Enlightenment of the heart is necessary for positive thinking and to develop wisdom. Thinking should take a positive bent as it develops. It originates from the heart and is maintained in the mind. Thinking can have a profound effect on our emotional and spiritual health.

Positive Emotions and Happiness

Positive emotions solve problems concerning personal growth and development. Experiencing positive emotions lead a person to develop a mature state of mind that equips him to cope with hard times ahead. Just as negative emotions provided strong evolutionary advantage for our ancestors by promoting self-protective responses to aversive circumstances, positive emotions help us to explore and understand our environments. The beneficial effect of positive emotions may be part of what drives us to explore, understand and develop ourselves and our worlds, changing ours as well as others' lives for the better. Thus, both negative and positive emotions are psychologically and adaptively worthy of themselves.

Fredrickson (2000) suggested that positive emotions could be cultivated indirectly by: (a) finding positive meaning within current circumstances; (b) finding positivity within adversity; (c) infusing ordinary events with meaning; and (d) effective problem solving (Baumgardner and Crothers, 2009; Fredrickson, 2003) both individually and collectively. Fredrickson (2001) proposed her 'Broaden and Build' theory which provides an overview of how positive emotions help build physical, psychological and social resources. She proposed that positive emotions broaden an individuals' momentary mindset, and by doing so these help an individual in building up enduring personal resources. Experiencing positive emotions regularly may keep one healthy by making the body better able to ward off infections, and decreasing the likelihood of getting sick. The capacity to experience positive emotion is a fundamental human strength central to the study of human flourishing (Fredrickson, 2001). Experiencing positive emotions leads to overlapping of people's feelings of self and other in the beginning of a self–other relationship. It is one effect of positive emotions' propensity to broaden peoples' momentary mindset. If, during experiencing positive emotions, self-boundaries expand and become more permeable, people might more readily see their oneness with others and think in terms of 'we' and 'us' instead of 'me' *versus* 'you'. It may enhance relationship and lead to greater relationship satisfaction (Waugh and Fredrickson, 2006). Happy people become happier through kindness. Happy people have more happy memories in daily life in terms of quantity and quality. If people become more conscious of kindness in daily life, they may increasingly wish to be kind to others, enjoying the grace of being kind and feel grateful when on the receiving end of kindness (Otake, Shimai, Tanaka-Matsumi, Otsui, and Fredrickson, 2006).

Happiness, a composite of life satisfaction, coping resources and positive emotions, predicts desirable life outcome in many domains. Empirical research findings indicate that happy people become more satisfied not simply because they feel better but because they develop resources for living well (Cohn, Fredrickson, Brown, Mikels and Conway, 2009). Positive emotions broaden the scope of attention and thought–action repertoire (Fredrickson and Branigan, 2005). Tugade and Fredrickson (2004) empirically demonstrated that positive emotions contribute to psychological and physical well-being via more effective coping. A number of studies have concluded that positive emotions and optimism can have a beneficial influence on health (Diener and Chan, 1984).

Positive emotions might signify optimal functioning. Fredrickson and Losada (2005) described a quantitative relationship between people's emotional experiences and their level of optimal functioning. It is well established that positive emotions lead to enhanced psychological and physical functioning. One suggestion by Mauss *et al.* (2011) was that a smile, e.g., which is a positive emotional behaviour accurately signals to others the individual's positive emotional state enhancing his social connectedness and thereby nurturing happiness and well-being. The study of positive emotions has traditionally not been given the same attention in research as negative emotions (Fredrickson, 2001). The term 'positive affect' is often used synonymously with positive emotions. Positive affect generally is defined as the extent to which an individual feels alert, enthusiastic and active (Watson, Clark and Tellegen, 1988). Frequent positive affect along with infrequent negative affect has been characterized as both necessary and sufficient condition to produce a state of happiness or affective well-being (Diener, Sandvik and Pavot, 1991).

Interpersonal Relationships and Happiness

Relationship among family members, friends and workmates make our life happy. Every relationship has to be nurtured well in order to get happiness. Relationships like marriage, kinship, close friendship, cooperation with acquaintances and involvement in religion and spiritual practices are all associated with lasting happiness.

Marriage: Marriage usually leads to a rapid increase in subjective well-being (SWB). So marriage changes the set point of SWB, although this change is not large. Married people are happier than unmarried people, lest they are divorced or separated (Myers, 2000). Ironically, the least happy people are the ones enduring unhappy marriages. The difference in happiness between married and unmarried women is the same as that of men. Both man and women reap the same benefits in terms of personal happiness from marriage. There are two reasons for the connection between happiness and marriage. One is that it is happy people who are more inclined to get married while the unhappy people do not because the latter find themselves unfit for marriage for various reasons. Another meaning is that marriage confers a range of benefits on people that make them happy. Marriage provides psychological and physical intimacy, a context to have children and build a home, a social role as a spouse and parent and a context to affirm identity and create posterity.

The National Opinion Research Centre surveyed over 35,000 Americans over the last 30 years. Results revealed that 40 per cent of married people called themselves very happy, while only 23 per cent of never-married adults made the same claim. This was true of every ethnic group studied. Marriage makes people happy. This finding suggests that married people are usually more outgoing and sociable than depressed people (Hoggard, 2005). For people who get married, their happiness levels reach to the peaks, but after a while, their happiness level returns to their pre-marriage days (Hoggard, 2005). Those who were married were considerably happier than those who were not. Least happy of all, on average, were those who had been married but were divorced, separated or widowed (Nettle, 2005).

A study conducted in Germany on 24,000 people yearly for up to 15 years showed that transition to marriage in an individual was associated with a substantial increase in happiness. Within 2 years, though, this hike generally melted away, and they were essentially back at baseline. Interestingly, the researchers observed considerable variation in the response to marriage. Some people, who responded with high increase in happiness in the short term, retained that increment for many years. On the other hand, some people, whose initial reaction to marriage was relatively weaken were actually less happy a couple of years later (Nettle, 2005).

In 2005, Martin conducted an international comparison to examine the association between marriage and happiness (Martin, 2005). He found that marriage was positively correlated with happiness in 16 out of 17 nations (the exception was Northern Ireland). Marriage appears to be a potential source of happiness. The key to successful relationships is to primarily ensure happiness within oneself. Many people end up in unhappy relationships because they seek a partner for emotional or financial security rather than as a long term soulmate, friend or lover. Many people settle for second best for fear of being alone or of having to go through the process again, and many marry far too young, pledging the entire life to someone before they have grown as an individual. It is important to remember that a person can never feel truly happy and fulfilled with someone else unless he can feel content, happy and secure with himself.

Friendship: Keeping few close confiding friends has been found to increase happiness and subjective well-being (Argyle, 2000, 2001). In a study of the happiest 10 per cent of a group of 222 college students, Diener and Seligman (2002) found that their most distinctive attribute was their rich and fulfilling social life. These students spent significant amount of their time socializing with friends and were rated by themselves and their friends as being outstanding in making and maintaining close friendship. Close friendships are probably associated with happiness due to the following reasons. First, people more often may select as friends and confide because they are more attractive companions than miserable people. They also help others more than depressed people who are self-focused and less altruistic. Second, close relationship meets the need for affiliation and so makes a person feel happy and satisfied. Third, close friendships provide social support. These research findings and insights on friendship from Evolutionary Psychology (Buss, 2000) have implications as to how a person can enhance his happiness through relationship with friends.

Acquaintances: *Life itself has become a competition that the chance of frustration or lost feeling among people has become more pronounced.* Against this background association with acquaintances is a potential source of solace and happiness and a way of avoiding unhappiness due to loss of status and inequalities which inevitably arises from (Axelrod, 1984; Buss, 2000). It follows therefore that to enhance a person's own sense of well-being, he should develop strategies for promoting cooperation with acquaintances.

Nature and Happiness

From the very origins of our species nature has shaped how we live and how we perceive the world in which we live. Nature has always been a source of awe, a source of survival, a source of pleasure and a source of fear. Nature has given us the sense of being a small speck in the universe and of sensing the existence of a supreme creator. As such our ancestors learned to respect, appease, make offerings and deify it. Many cultures have looked on the sun and moon as deities, or seen features of the natural landscape as the abode of various deities. Mountains and waterfalls, animals and trees have either become gods and spirits or the home of such beings. Across all cultures, places and sites are considered sacred based on their unusual soothing or healing power.

Human beings are both happier and healthier as a result of interacting with nature. Kaplan and Kaplan (1989) concluded, 'People with access to nearby natural settings have been found to be healthier overall than other individuals. The long term, indirect impacts (of "nearby nature") also include increased levels of satisfaction with one's home, one's job and with life in general' (p. 173). In pointing out the relationship between nature and happiness, Burns (2009) believes that Bhutan illustrates this relationship in a subtle yet important manner. That is, more than supporting our survival, nature has a deep impact on our happiness and psychological well-being. Contact with nature can provide a buffer against emotional distress, thus serving a preventative role in the area of mental health. Generally, the happier we are, the greater our range of positive emotional responses, and the less likely we are to slip into clinically unhealthy states of sadness, anxiety or anger (Fredrickson, 2005, 2008).

Extraversion and Happiness

Eysenck (1983) noted, 'Happiness is a thing called stable extraversion…the positive effect in happiness seemed to be related to easy sociability, with a natural, pleasant interaction with other people…then it only makes sense that happiness can be associated with extraversion. Similarly, if worries and anxieties make up negative affect in happiness, it can easily be seen that instability and neuroticism are also connected to unhappiness' (p. 67).

There are a number of factors which offer partial explanations for the link between extraversion and happiness (Diener, Suh, Lucas and Smith, 1999). Extraverts may better fit in with the social environment, which requires them to be involved in frequent social interactions. So they find themselves frequently in situations which meet their needs for socializing and thus are happier. There is good evidence that extraverts respond with greater happiness to stimuli designed to induce positive moods. There is also evidence that extraversion and neuroticism predispose people to experience more positive and negative events, respectively. So, if a person has a high level of extraversion, he is more likely to experience positive events and, therefore, experience more happiness. Part of the evidence for there being a consistent happiness trait is that measures of happiness are stable over a number of years. In one study, happiness was successfully predicted from associated traits like extraversion over an interval of 10–17 years

(Costa, McCrae and Norris, 1981). The strongest correlate of happiness is extraversion—if extraversion is divided into sociability and impulsiveness components it is sociability that predicts happiness. Extraversion particularly predicts positive moods during social interaction.

Headey and Wearing (1992) carried out a 4-stage panel study, at 2-year intervals, of 649 Australians. They found that extraversion was a cause of later positive effect and did so by generating favourable life-events, especially in the spheres of friendship and work. So while friendship and work were causes of happiness, they were themselves, in part, due to extraverts seeking out and generating positive activities. Furnham (1981) found that extraverts choose social and physical activities kinds of situation.

Furnham (1981) found that extraverts choose social and physical activities kinds of situation. The happiness of extraverts is partly explained by their choice of enjoyable social activities. In addition, extraverts possess certain skills. We found that extraverts score high in assertiveness, and in a longitudinal study this predicted later happiness and partly explained the happiness of extraverts. This shows how extraverts cause happiness via the route: extraversion-assertiveness-happiness (Argyle and Lu, 1990). In a similar study we found that certain aspects of cooperativeness also predicted happiness, particularly enjoying engaging in joint leisure with friends. Chamorro-Premuzic, Bennett and Furnham (2007) found that four of the Big Five factors, viz., stability, extraversion, conscientiousness, and agreeableness were positively correlated with both happiness and emotional intelligence, which explained18 per cent of unique variance (over and above age and the Big Five) unhappiness.

Social Support and Happiness

Social support is believed to be an important factor in enacting happiness changes. Interpersonal support can aid an individual both in initiating a potential happiness—increasing activity and in maintaining it. Performing an intentional activity as a group or with the support of close others is likely to promote greater and more sustained happiness change than 'bowling alone' (Putnam, 2000). Social support increases subjective well-being as it helps people to derive happiness from kinship network (Argyle, 2001; Buss, 2000). There are certain things that can be done to enhance the contributions of kinship on happiness. Keep in regular contact with members of the family. Plan a lifestyle that allows closer physical contact with the family. This planning refers to both the short run and the long run. Maintaining contact with the family members increases social support and this brings not only happiness but also enhances immune system functioning.

Self-esteem and Happiness

There are several sources (e.g., social circumstances or events) that produce happiness, one important source is relatively stable personal traits and dispositions such as self-esteem (Diener, Suh, Lucas and Smith, 1999; McCrae and Costa, 1991). A strong relationship exists between self-esteem and happiness (Baumeister, Campbell, Krueger and Vohs, 2003), i.e., high self-esteem leads to greater happiness. The pursuit of happiness is one of the overarching goals in life for most people, and feeling happy is one form of positive outcome.

Before we present findings, we must acknowledge that studies of happiness and self-esteem seem almost inevitably to rely on self-reports. Hence, one might well expect that people with high self-esteem claim to be happier than other people. A major international study of self-esteem and happiness was reported by Diener and Diener (1995). The data came from more than 13,000 college students from 49 different universities, 31 countries and five continents. High self-esteem emerged as the strongest of several predictors of life satisfaction overall. Self-esteem and happiness were found substantially interrelated. The correlation between self-esteem and happiness (life satisfaction) varied somewhat across countries. In particular, it was stronger in individualistic countries than in collectivistic ones. Diener and Diener (1995) proposed that in individualistic countries, people are socialized to attend to their own internal attributes, and so these become important in determining overall happiness, whereas collectivistic cultures encourage people to focus on groups and relationships.

A meta-analysis of the relation between 137 personality traits and happiness (subjective well-being) was published by DeNeve and Cooper (1998). They found that 'private collective self-esteem' was one of the strongest predictors of happiness. Their results provide further evidence that self-esteem is consistently correlated with happiness. In a sample of 406 young people (aged between 14–28 years), Furnham and Cheng (2000) measured a number of potential correlates of happiness and they reported that self-esteem was the most dominant and powerful predictor of happiness. Shackelford (2001) examined the happiness of young to middle-aged couples (aged between 17–41 years) who had been married within the past year, found that for both husbands and wives, self-esteem was significantly correlated with happiness in the form of global, sexual and emotional satisfaction. Lyubomirsky and Lepper (2002) obtained data from more than 600 older adults, aged 51–95 years. Once again, happiness and self-esteem were found to be highly correlated (r = .58). In this study, self-esteem was found to be more strongly correlated to hopelessness, optimism and sense of mastery, whereas happiness was more strongly correlated to energy level, overall health, loneliness, mood and emotion, and purpose in life.

In Diener and Diener's (1995) work, self-esteem and happiness had very different patterns of correlations with other predictor variables, which supports the conclusion that happiness and self-esteem are in fact distinct constructs that can be measured separately despite their relatively high links. Martin's (2005) research indicates that global self-esteem has a stronger influence on children's overall happiness, whereas academic self-esteem has a stronger influence on how children behave in the classroom. According to him, academic self-esteem can contribute to global self-esteem, and hence to overall happiness.

Taken together, these findings uniformly indicate that self-esteem and happiness are strongly interrelated. Further research with longitudinal designs (or other means of testing causal relationships), controls for third variables, and, if possible, alternatives to self-report measures would greatly strengthen the case. The link between self-esteem and happiness is strong. People with high self-esteem are significantly, substantially happier than other people. They are also less likely to be depressed, either in general or specifically in response to stressful, traumatic events.

Conclusion

The factors discussed above suggest that this is a practical way to look at life and happiness. Happiness can be defined as composite of *Health, Attributional style, Positive thinking, Positive emotions, Interpersonal relationships, Nature, Extraversion, Social support, and Self-esteem.* Considering happiness as an acronym, this model can be used to develop a broad theory that explains the concept of happiness.

The role of positive emotions will continue to be a most interesting aspect of future happiness researches. To make progress, we will need to better objectify the matter of greatest individual concern—the dynamics of affective experience—at more than a superficial level. Understanding the nature of human happiness will depend critically on our ability to work out the deeply unified emotional nature of the brain, the body and the mind.

References

Argyle, M. (1987). *The Psychology of Happiness*. London: Methuen.

Argyle, M. (1997). Is Happiness a Cause of Health? *Psychology and Health*, *12*(6), 769–81.

Argyle, M. (2000). *Psychology and Religion: An Introduction*. London: Routledge.

Argyle, M. (2001). *The Psychology of Happiness*. (2ndEd.). London: Routledge.

Argyle, M., and Lu, L. (1990). The Happiness of Extraverts. *Personality and Individual Differences, 11*(10), 1011–17.

Argyle, M., Martin, M., and Crossland, J. (1989). Happiness as a Function of Personality and Social Encounters. In J. P. Forgasand J. M. Innes (Eds.), *Recent Advances in Social Psychology: An International Perspective* (pp. 189–203). North-Holland: Elsevier.

Axelrod, R. (1984). *The Evolution of Co-operation*. New York: Basic Books.

Baumgardner, S. R., and Crothers, M. K. (2009). *Positive Psychology.* New Delhi: Pearson Education.

Baumister, R. F., Campbell, J. D., Krueger, J. I., Vohs, K. D. (2003). Does High Self-esteem Cause Better Performance, Interpersonal Success, Happiness or Healthier Lifestyles? *Psychological Sciences in the Public Interest*, *4*(1), 1–44.

Burns, G. W. (2009). Gross Natural Happiness: Can We Have Both Psychological and Ecological Well-being? In U. Karma and D. Penjore, (Eds.), *Gross National Happiness. Practice and Measurement* (pp.127–148). Thimpu, Bhutan: The Center for Bhutan Studies.

Buss, D. (2000). The Evolution of Happiness. *American Psychologist, 55*(1), 15–23.

Chamorro-Premuzic, R., Bennett, E., and Furnham, A. (2007). The Happy Personality: Mediating Role of Trait Emotional Intelligence. *Personality and Individual Differences, 42*(8), 1633–39.

Cohn, M. A., Fredrickson, B. L., Brown, S. L., Mikels, J. A., and Conway, A. M. (2009). Happiness Unpacked: Positive Emotion Increase Life Satisfaction by Building Resilience. *Emotion, 9*(3), 361–68.

Costa, P. T., McCrae, R. R., and Norris, A. H. (1981). Personal Adjustment to Aging: Longitudinal Prediction from Neuroticism and Extraversion. *Journal of Gerontology, 36*(1), 78–85.

Danner, D. D., Snowdon, D. A., and Friesen, W. V. (2001). Positive Emotions in Early Life and Longevity: Findings From the Nun Study. *Journal of Personality and Social Psychology, 80*(5), 804–13.

Davidson, R. J., Kabat-Zinn, J., Schumacher, J., Rosenkranz, M., Muller, D., and Santorelli, S. F. (2003). Alterations in Brain and Immune Function Produced by Mindfulness Meditation. *Psychosomatic Medicine, 65*(4), 564–70.

DeNeve, K. M., and Cooper, H. (1998). The Happy Personality: A Meta-analysis of 137 Personality Traits and Subjective Well-being. *Psychological Bulletin, 124*(2), 197–229.

Diener, E. (1984). Subjective Well-being. *Psychological Bulletin, 95*(3), 542–75.

Diener, E. (2008). *Happiness: Unlocking the Mysteries of Psychological Wealth*. Malden, MA: Blackwell Press.

Diener, E., and Biswas-Diener, R. (2008). *Happiness: Unlocking the Mysteries of Psychological Wealth.* Malden, MA: Blackwell Publishing.

Diener, E., and Chan, M. Y. (1984). Happy People Live Longer: Subjective Well-being Contributes to Health and Longevity. *Applied Psychology: Health and Well-being, 3*(1), 1–43.

Diener, E., and Diener, M. (1995). Cross-cultural Correlates of Life Satisfaction and Self-esteem. *Journal of Personality and Social Psychology, 68*(4), 653–63.

Diener, E., and Oishi, S. (2004). Are Scandinavians Happier Than Asians? Issues in Comparing Nations on Subjective Well-being. In F. Columbus (Ed.), *Asian Economic and Political Issues* (Vol. 10, pp. 1–25). Hauppauge, NY: Nova Science Publishers.

Diener, E., Sandvik, E., and Pavot, W. (1991). Happiness is the Frequency, Not the Intensity, of Positive Versus Negative Affect. In F. Strack, M. Argyle, and N. Schwarz (Eds.), *Subjective Well-being: An Interdisciplinary Perspective* (pp. 119–39). New York: Pergamon.

Diener, E., and Seligman, M. (2002). Very Happy People. *Psychological Science, 13*(1), 81–84.

Diener, E., Suh, E. M., Lucas, R. E., and Smith, H. L. (1999). Subjective Well-being: Three Decades of Progress. *Psychological Bulletin, 125*(2), 276–302.

Eysenck, H. J. (1983). *I do Your Guide to a Happy Marriage*. London: Century.

Fredrickson, B. L. (2000). Extracting Meaning From Past Affective Experiences: The Importance of Peaks, Ends and Specific Emotions. *Cognition and Emotion, 14*(4), 577–606.

Fredrickson, B. L. (2001). The Role of Positive Emotions in Positive Psychology: The Broaden-and-Build Theory of Positive Emotions. *American Psychologist, 56*(3), 218–26.

Fredrickson, B. L. (2003). The Value of Positive Emotions. *American Scientist, 91*(4),330–35.

Fredrickson, B. L. (2005). The Broaden-and-build Theory of Positive Emotions. In F. A. Huppert, N. Baylis, and B. Keverne (Eds.), *The Science of Well-being* (pp. 217–38). Oxford: Oxford University Press.

Fredrickson, B. L. (2008). Promoting Positive Affect. In M. Eid, and R. J. Larsen, (Eds.), *The Science of Subjective Well-being* (pp. 449–68). New York: Guilford Press.

Fredrickson, B. L., and Branigan, C. (2005). Positive Emotions Broadens the Scope of Attention and Thought-action Repertories, *Cognition and Emotion, 19*(3), 313–32.

Fredrickson, B. L., and Losada, M. F. (2005). Positive Affect and the Complex: Dynamics of Human Flourishing. *American Psychologist, 60*(7), 678–84.

Furnham, A. (1981). Personality and Activity Preference. *British Journal of Social Psychology, 20*(1), 57–68.

Furnham, A., and Cheng, H. (2000). Lay Theories of Happiness. *Journal of Happiness Studies, 1*(2), 227–46.

Gilbert, D. (2006). *Stumbling on Happiness*. New York: Alfred A. Knopf.

Graves, D. (2000). The Obsessive Pursuit of Health and Happiness. *British Medical Journal, 321*(72–76), 1572–76.

Headey, B. W. and Wearing, A. J. (1992). *Understanding Happiness: A Theory of Subjective Well-being*. Melbourne: Longman Cheshire.

Hoggard, L. (2005). *How To Be Happy*. London: B.B.C. Books.

James, W. (1902/1958). *The Varieties of Religious Experience.* New York: Mentor Books.

Kaplan, R., and Kaplan, S. (1989). *The Experience of Nature: Apsychological Perspective.* Cambridge: Cambridge University Press.

Kozma, A., and Stones, M. J. (1983). Predictors of Happiness. *Journal of Gerontology, 38*(5), 626–28.

Larsen, R., Diener, E., and Emmons, R. (1985). An Evaluation of Subjective Well-being Measures. *Social Indicators Research, 17*(1), 1–17.

Lyubomirsky, S. (2001). Why are Some People Happier Than Other? The Role of Cognitive and Motivational Process in Well-being. *American Psychologist, 56*(3), 239–49.

Lyubomirsky, S., and Lepper, H. S. (1999). A Measure of Subjective Happiness: Preliminary Reliability and Construct Validation. *Social Indicator Research, 46*(2), 137–55.

Lyubomirsky, S., and Lepper, H.S. (2002). *What are the Differences Between Happiness and Self-esteem?* Unpublished Manuscript, University of California, Riverside.

Lyubomirsky, S., and Ross, L. (1999). Changes in Attractiveness of Elected, Rejected, and Precluded Alternativeness: A Comparison of Happy and Unhappy Individuals. *Journal of Personality and Social Psychology, 76*(6), 988–1007.

Macquarrie, J., and Childress, J. (2001). *A New Dictionary of Christian Ethics*. UK: Scum Press.

Martin, P. (2005). *Making Happy People.* London: Harper Perennial.

Mauss, I. B., Shallcross, A. J., Troy, A. S., John, O. P., Ferrer, E., Wilhelm, F. H., and Gross, J. J. (2011). Don't Hide your Happiness! Positive Emotion Dissociations, Social Connectedness, and Psychological Function. *Journal of Personality and Social Psychology, 100*(4), 738–48.

McCrae, R. R., and Costa, P. T. (1991). Adding Liebe und Arbeit: The Full Five-factor Model and Well-being. *Personality and Social Psychology Bulletin, 17*(2), 227–32.

Myers, D. (2000). The Funds, Friends and Faith of Happy People. *American Psychologist 55*(1), 56–67.

Nettle, D. (2005). *Happiness.* Oxford: Oxford University Press.

Okun, M. A., Stock, W. A., Haring, M. J., and Witter, R. A. (1984). Health and Subjective Well-being. *International Journal of Aging and Human Development, 19*(2),111–32.

Otake, K., Shimai, S., Tanaka-Matsumi, J., Otsui, K., and Fredrickson, B. L. (2006). Happy People Become Happier to Kindness: A Counting Kindness Intervention, *Journal of Happiness Studies, 7*(3), 361–65.

Pressman, S. D., and Cohen, S. (2012). Positive Emotion Word Use and Longevity in Famous Deceased Psychologists. *Health Psychology, 31*(3), 297–305.

Putnam, R. (2000). *Bowling Alone–The Collapse and Revival of American Community*. New York: Simon and Schuster.

Ryan, R. M., and Deci, E. L. (2001). On Happiness and Human Potentials. A Review of Research on Hedonic and Eudaimonic Well-being. *Annual Review of Psychology, 52*(1), 141–66.

Ryff, C.D., and Keyes, C. L. M. (1995). The Structure of Psychological Well-being Revisited. *Journal of Personality and Social Psychology 69*(4), 719–27.

Seligman, M. E. P. (2002). *Authentic Happiness: Using the New Positive Psychology to Realize Your Potential for Lasting Fulfillment*. New York: Free Press.

Shackelford, T. K. (2001). Self-esteem in Marriage. *Personality and Individual Differences, 30*(3), 371–90.

Stones, M. J., and Kozma, A. (1980). Issues Relating to the Usage and Conceptualization of Mental Health Constructs Employed by Gerontologists. *International Journal of Aging and Human Development, 11*(4), 269–82.

Stones, M. J., and Kozma, A. (1985). Structural Relationships among Happiness Scales: A Second Order Factorial Study. *Social Indicator Research*, *17*(1), 19–28.

Tugade, M. M., and Fredrickson, B. L. (2004). Resilient Individuals Use Positive Emotion to Bounce Back from Negative Emotional Experiences. *Journal of Personality and Social Psychology*, *86*(2), 320–33.

Veenhoven, R. (1989). Introduction. In R. Veenhoven (Ed.), *How Harmful is Happiness?* (pp. 1–6). Rotterdam, the Netherlands: Universitaire Pers Rotterdam.

Wallace, B. A., and Shapiro, S. L. (2006). Mental Balance and Well-being: Building Bridges between Buddhism and Western Psychology. *American Psychologist, 61*(7), 690–701.

Watson, D. (1988). Intra-individual and Inter-individual Analyses of Positive and Negative Affect: Their Relation to Health Complaint, Perceived Stress and Daily Activities. *Journal of Personality and Social Psychology*, *54*(6), 1020–30.

Watson, D., Clark, L. A., and Tellegen, A. (1988). Development and Validation of Brief Measures of Positive and Negative Affect: The PANAS Scales. *Journal of Personality and Social Psychology*, *54*(6), 1063–70.

Waugh, C. E., and Fredrickson, B. L. (2006). Nice to Know You: Positive Emotions, Self-other Overlap, and Complex Understanding in the Formation of New Relationship. *Journal of Positive Psychology, 1*(2), 93–106.

White, N. (2006). *A Brief History of Happiness*. UK: Blackwell Publishing.

Part 2

Applications in Work, Health and Well-being

Chapter 7

Psychological Capital as a Catalyst to Positive Organizations

Papri Nath

Indian Institute of Management, Tiruchirappalli, Tamil Nadu

Rabindra Kumar Pradhan

Indian Institute of Technology, Kharagpur, West Bengal

Absract The study of positive organizational behaviour has discovered constructs like hope, self-efficacy, optimism and resilience. These constructs are not conceptually identical. These constructs differ in their measurement and their combination produces a higher order construct. This higher order construct is termed as psychological capital or Psycap. The impact of Psycap is greater than the sum of its individual components. Psycap can be developed and improved through intervention and proactive management techniques at the workplace. This developmental nature of Psycap has implications in enhancing organizational effectiveness and employee well-being. The overall intent of this chapter is to discuss how Psycap acts as a catalyst in the development of positive organizations.

The recent global economic slowdown has generated a host of issues at the workplace. Some major ones are unemployment, job insecurity, hopelessness and general pessimism. In order to counter this situation, a positive mindset needs to be promoted. Seminal research by Seligman and Csikszentmihalyi (2000) and Luthans (2002b) encouraged researchers to adopt a positive approach. Strumpfer (2005) and Roberts (2006) also advocate adopting a positive approach in organizations. Positive approach takes a 'strengths'-based perspective which may aim to understand the dynamics of 'optimal human performance' during challenges and adversities. Following the positive approach advocated by Seligman and Csikszentmihalyi (2000), Luthans and colleagues (Avolio and Luthans, 2006; Luthans, Avey, Avolio, Norman and Combs, 2006; Luthans, Youssef and Avolio, 2007) have brought a positive approach to organizational behaviour. This approach has been termed as positive organizational behaviour (POB). POB refers to a positive or strengths-based approach with focus on identifying the desired behaviour at workplace. It forms a bridge between popular literature and scientific knowledge (Luthans, 2002a). This is a critical issue for researchers and HR practitioners who play important role as change agents at both organizational and societal level. The development of POB has significant impact on employee outputs, specifically on an organizations' bottom line. Luthans and colleagues (2007) acknowledge that this impact on the bottom line is relevant to both public and private organizations.

The domain of POB focuses on identifying and understanding individual strengths and how they grow and thrive (Luthans, Norman, Avolio, and Avey, 2008a). A major contribution of POB movement in understanding human strength is the concept of psychological capital or PsyCap. Researches taking a POB perspective have demonstrated that PsyCap positively relates to desirable work outcomes (Luthans, Avolio, Avey, and Norman, 2007; Luthans *et al.*, 2008a), and negatively with undesirable ones (Avey, Luthans and Jensen, 2009; Avey, Luthans, and Youssef, 2010b; Larson and Luthans, 2006; Luthans *et al.*, 2008a).

Psychological capital is a label and is given to construct which best fits the four basic criteria to be accepted as a POB capacity. The four basic criteria for a POB concept are: (a) the uniqueness of the concept; (b) valid measure; (c) open to development; and (d) capable of performance improvement at the workplace (Luthans, 2002a, 2002b, 2002c). Luthans in his article on POB (Luthans and Youssef, 2004) listed four positive psychological capacities which best fit the criteria of POB. They are *self-efficacy*, *hope, resiliency* and *optimism* (Luthans and Youssef, 2004).

Luthans and Youssef (2007) have proposed that these positive capacities form a higher order construct which has the capacity to determine the direction of our near and distant future behaviour. This argument is derived from the theoretical underpinnings in psychological resource theories (Hobfoll, 2002). Resource theories assert that some psychological variables such as hope, optimism, resiliency, and self-efficacy are best understood and treated as manifestations of a larger underlying 'resource' which is evident in each capacity. They could be described as a set of capacities (Avey, Patera, and West, 2006). These four positive capacities together are termed as positive psychological capital or PsyCap (Luthans and Youssef, 2007).

Youssef and Luthans (2007) have pointed out the importance of positive capacities by differentiating between positive state and trait. They explain that positive states are momentary and highly changeable. They represent one end of the state-trait continuum. State-like positive capacities like hope, optimism, resiliency and self-efficacy come next in this continuum which are more stable and are open to change and development. And at the other end of the continuum fall traits which are unlikely to change. According to Luthans, Youssef and Avolio (2007) state-like positive capacities exhibit lower though significant test-retest correlations than do trait-like personality (Youssef and Luthans, 2007). They found empirical evidence that these capacities can be developed within short, highly focused training interventions (Luthans, Youssef and Avolio, 2007). These evidences make one thing clear that state-like positive capacities are stable forms of momentary positive states and could be changed and developed via training exercises. Thus, positive capacities might act as catalyst in creating positive work environment.

Self-efficacy: I can!

The concept of self-efficacy as a POB capacity comes from Bandura's (1994) concept of self-efficacy. The term efficacy is interchangeably used with the term 'confidence' in the POB literature (Luthans, 2002b). The self-efficacy construct rest upon a long historical thinking related to the sense of personal control (Snyder and Lopez, 2008). Bandura (1997) defined self-efficacy as 'peoples beliefs in their capabilities to produce desired effects by their own actions'. Based on an examination of which particular behaviour will produce a particular outcome (outcome expectancies) (Bandura, 1977, 1982, 1989; Maddux and Volkmann, 2010), an individual analyses own capacity to perform a particular behaviour that will lead to a desired outcome (efficacy expectancies) (Snyder *et al.*, 1991). According to Bandura (1977), outcome expectancies are viewed as far less important than efficacy expectancies. Self-efficacy thoughts are considered to be most important cognitive step before people launch goal directed behaviour (Snyder and Lopez, 2008). In POB literature self-efficacy is defined as 'an individual's convictions (or confidence) about his or her abilities to mobilize the motivation, cognitive resources, and courses of action needed to successfully execute a specific task within a given context' (Stajkovic and Luthans, 1998b, p. 66). Self-efficacy defined here is specific to task and context (Luthans, 2002a, 2002b; Stajkovic and Luthans, 1998a) also termed as specific self-efficacy.

Luthans (2002a) follows the work by Bandura (1997) to provide justification as to how self-efficacy qualifies as a POB capacity. Bandura (1997) was the first to take a positive strength- based approach and propose that self-efficacy facilitates to select and structure ones' environment in ways that lead to a successful course (Snyder and Lopez, 2008, p. 177). Taking evidence from such seminal work, Luthans (2002b) proposed that work on self-efficacy is a positive approach and is open to development—this is backed by rich theory and considerable research support. Self-efficacy is a learned human pattern of thinking than an inherent trait. Self-efficacy is primarily conceptualized (Bandura, 1997) and measured (e.g., Maurer and Pierce, 1998; Parker, 1998) as a state. Its state-like nature is manifested not only in its developmental nature over time but also in its domain specificity. Having efficacy in one domain is not necessarily transferable to other domains, whereas lacking efficacy in some contexts does not ensure being efficacious in others (Bandura, 1997 as cited in Luthans and Youssef, 2007). Bandura (1997) suggested that any measurement of self-efficacy should be carefully tied to a given performance situation. High self-efficacy has been found to predict better academic performance, continuation in smoking cessation treatment, and adoption of exercise and diet regimes (Snyder and Lopez, 2002). Thus, self-efficacy successfully fulfils all the criteria set for a POB capacity.

Optimism: Positive Attribution Style

Optimism as a component of PsyCap is borrowed from the positive psychology movement. Optimism is a positive explanatory/attribution style that explains positive events through personal, permanent and pervasive causes and negative events through external, temporary, and situation-specific ones. On the other hand, pessimistic explanatory style attributes positive events to external, temporary and situation-specific causes whereas it attributes negative events to internal, permanent and pervasive ones (Peterson, 2000; Seligman, 1998). The differences in attributing causes to positive and negative events, optimists build positive expectancies that facilitate their goal pursuit and approach-coping behaviour. In contrast, pessimists are restricted by self-doubt and negative expectancies (Carver and Scheier, 2002).

The concept of explanatory/attribution style emerged from the classic work on learned helplessness model done by Seligman (1972). The model proposes that after experiencing a series of uncontrollable adverse situations, an individual becomes helpless. The reason is that the individual cannot see any connection between his actions and the expected outcome. This develops an expectation that future actions will also be unrelated to outcomes (Peterson, 2000). The explanatory style has been linked to helplessness model for understanding as to how causal attributions of situations make people find connection between their action and its outcomes. These attributions determine how they respond to the situation. When they adopt an optimistic style it develops a sense of control over the situation which facilitates adaptive response.

Optimistic explanatory style can be learned and developed (Seligman, 1998) in the same way as helplessness through recognized approaches such as leniency for the past, appreciation for the present and opportunity seeking for future (Carver and Schneider, 2001 as cited in Youssef and Luthans, 2007). Thus, optimism as a POB capacity is open to development, could be validly measured and contributes in performance improvement at the workplace.

Hope: Where There is a Will There is a Way

Hope is the most unique POB capacity (Luthans, 2002a, 2002b, 2002c). Unlike other constructs of POB, hope was not found in the organizational behaviour literature. Researchers have defined hope as 'a cognitive set that is based on a reciprocally derived sense of success of two aspects—willpower and way power' (Snyder *et al.*, 1991; Snyder and Lopez, 2008). This definition follows the goal-setting framework, which highlights two elements of hope– agency (willpower, goal- directed energy) and pathways (way power, the perceived capacity to find routes to desired goals). Snyder (2002) explained that hope's agency component provides the determination to achieve goals, whereas pathways component promotes the creation of alternative paths to replace those that may have been blocked in the process of pursuing those goals (Youssef and Luthans, 2007). Thus, hope is defined as a cognitive set that is based on a reciprocally derived sense of successful agency and pathways (Luthans, 2002b; Snyder *et al.*, 1991).

Snyder, Feldman, Taylor, Schroeder and Adams (2000) described hope as being a trait but also emphasized that there is a more transitory form of hope that can be developed and managed. His theory notes that fluctuations in state hope serve as the variable based upon which the more enduring trait hope is built (Luthans and Jensen, 2002). This malleability of hope presents the potential to build on the existing strengths of individuals and enhance agency and pathways thinking (Luthans and Jensen, 2002). Snyder (2000) suggested that individuals who are not able to adapt with rapid technological advancements, could be encouraged to recall personal achievements of their own life and identify hopeful stories. Such techniques can help to recognize one's past ability to preserve and deal with obstacles which develop hope (Luthans and Jensen, 2002). Snyder, Sympson, Ybasco, Borders, Babyak and Higgins (1996) have developed a valid scale to capture the state form of hope. Therefore, hope is a unique POB capacity (Luthans, 2002b), as it meets the criteria of being state-like, open to development, and has a valid measure (Snyder *et al.*, 1991).

Resiliency: Bouncing Back

The concept of resiliency has its origin in clinical psychology where it is defined as class of phenomenon characterized by patterns of positive adaptation in the context of significant adversity or risk (Masten, 2001; Masten and Reed, 2002). Earlier it was considered as an extraordinary trait possessed by a few people. But now it considered as an 'everyday magic of the ordinary'. This characteristic of ordinariness makes resiliency significant in today's workplace (Luthans, 2002b).

Experiments by Fredrickson and colleagues suggest that resiliency is a trait which predicts both psychological well-being and physiological recovery (Fredrickson, 2003; Tugade and Fredrickson, 2004). A significant number of evidences suggest that resiliency also has a state-like form which is open to development (Coutu, 2002). Resiliency as a state-like capacity is defined as the capability to cope successfully during significant change, adversity, or risk (Luthans, 2002b).

Unlike the traditional approach, resiliency is not only concerned with reacting towards stress and adversity but involves proactive management of resources to deal with future adversities (Fredrickson, 1998, 2001). As noted by Masten (2001) these resources could be attachment, self-regulation, relations to competent and caring support, and motivation to be effective in the environment. Thus, as defined by Luthans (2002b) resiliency is defined as a positive psychological capacity to bounce back not only from negative situations but also from positive changes like progress, increased responsibility, etc.

The concept of resiliency is new and unique to the field of organizational behaviour. The concept of resilience has mainly been studied in context of stress, specifically *eustress* (i.e., positive stress). Besides this only a few attempts have been made to study resiliency at workplace and that too only at the organizational level (Doe, 1994; Home and Orr, 1998; Mallak, 1998). Therefore, resiliency qualifies the criteria as a POB concept, because it has a state-like form and is open to development (Carver, 1998; Dyer and McGuinness, 1996) and is also unique and measurable (Block and Kremen, 1996).

Taking all the shared characteristics of the components of PsyCap, Luthans and colleagues defined Psychological capital as 'an individual's positive psychological state of development that is characterized by: (1) having confidence (self-efficacy) to take on, and put in the necessary effort to succeed at challenging tasks; (2) making a positive attribution (optimism) about succeeding now and in the future; (3) persevering toward goals and, when necessary, redirecting paths to goals (hope) in order to succeed; and (4) when beset by problems and adversity, sustaining and bouncing back and even beyond (resiliency) to attain success' (Luthans, Youssef and Avolio, 2007, p. 3).

Psychological Capital as a Catalyst to Positive Organizations

PsyCap as a higher order construct incorporates the mechanism(s) that the above four discriminant constructs share in common (Avey, Wernsing, and Luthans, 2008; Avey, Reichard, Luthans, and Mhatre, 2011). As a result, the impact of PsyCap is greater than the sum of its individual components or capacities. PsyCap as a positive resource acts as a catalyst in enhancing positive job outcomes and reducing negative job outcomes at work, creating positive organizations.

PsyCap and Job Performance

PsyCap is positively related to performance (Hodges, 2010). Research found significant positive correlation between employees PsyCap and performance outcome of relative merit-based salary (Luthans, Avolio, Walumbwa, and Li, 2005). It is also reported that high PsyCap employees exhibited more innovative behaviour by their supervisors, than employees with lower PsyCap (Abbas and Raja, 2011). A study on the South African Board of People Practice demonstrated that PsyCap is an essential factor for HR professionals to promote employee performance (du Plessis and Barkhuizen, 2012). A meta-analysis indicated significant positive relationship of PsyCap with multiple measures of job performance (Avey *et al.*, 2011).

PsyCap and Job Commitment

Employees with high PsyCap are more committed towards their organization (Larson and Luthans, 2006; Luthans *et al.*, 2008a). A study in healthcare industry found significant positive relationship between PsyCap and commitment towards mission, values and goals of the hospital (Luthans and Jensen, 2005). A large cross-sectional study indicated that PsyCap was the greatest predictor of affective organizational commitment as compared to personality traits like core self-evaluations, conscientiousness and extraversion (Youssef and Luthans, 2007). A developmental study on soldiers of the US Army Reserves found hope and optimism to be related with affective commitment (Bressler, 2010).

PsyCap and Job Satisfaction

PsyCap has been found to be significantly related with job satisfaction (Avey *et al.*, 2010b; Avey *et al.*, 2008; Larson and Luthans, 2006). PsyCap is reported to be a better predictor of satisfaction than personality traits like conscientiousness and extraversion (Luthans *et al.*, 2007c). A study found that job satisfaction has positive relationship with resilience, hope and optimism (Cetin, 2011). According to a recent study PsyCap has direct and positive influence on job satisfaction of university teachers (Chang, Chen, Lin and Huang, 2013).

PsyCap and Physical Health

PsyCap and its dimensions have impact on physical health of an individual. Hope is reported to promote health related goals including improved pain management, prevention of ill health and safer sexual practices (Snyder, 2002). Self-efficacy increases health-related behaviours and decreases unhealthy ones; moreover helps to maintain these changes (Raghunathan and Trope, 2002). Optimism leads to a broad range of positive outcomes, including perseverance, achievement, health and happiness (Luthans, Lebsack and Lebsack, 2008b; Scioli, Chamberlin, Samor, Lapointe, Campbell, Macleod and Maclenon, 1997). Psychological resilience is found to be a strong predictor of perceived physical health (Nath and Pradhan, 2012).

PsyCap and Employee Well-being

Psychological capital of employees is related to their well-being over time (Avey, Luthans, Smith and Palmer, 2010a). A cross-sectional survey demonstrated that PsyCap was negatively associated with depressive symptoms (Liu, Chang, Fu, Wang and Wang, 2012). Psychological capital is reported to partially mediate the impact of unemployment on well-being (Cole, Daly and Mak, 2009). A study on nurses revealed that PsyCap has positive impact on the quality of work life of the employees (Mortazavi, Yazdi and Amini, 2012). A two-wave study demonstrated that PsyCap was significantly and positively related with participants' well-being at work (higher levels of job satisfaction and physical/psychological well-being) (Oi Ling, 2013).

Impact of PsyCap on Job Absenteeism and Turnover Intentions

PsyCap has interesting influences on absenteeism at work. Few meta-analyses have shown that attitude variables like satisfaction and commitment cannot explain the significant variance in absenteeism (Bycio, Hackett and Allen, 1995; Harrison and Martocchio, 1998; Harrison, Newman and Roth, 2006). Researches show that PsyCap has a positive impact on both voluntary and involuntary absenteeism (Avey *et al.*, 2006). A number of studies also report that PsyCap is negatively associated with turnover intentions (Avey *et al.*, 2009; Avey *et al.*, 2010b). Therefore, PsyCap extends the understanding and management of absenteeism (Avey *et al.*, 2006) and subsequently turnover intentions at the workplace.

PsyCap and Job Stress

PsyCap acts as personal coping resource which forms a buffer between stress and burnout at the workplace (Herbert, 2011). It is recommended that development of PsyCap is important to reduce occupational stress among physicians (Liu *et al.*, 2012). A study on a heterogeneous sample of working adults considered PsyCap as the positive strength for combating workplace stress (Avey *et al.*, 2009). Employees with high psychological capital reported less job stress as compared to employees with low PsyCap (Abbas and Raja, 2011). PsyCap is also reported to be a strong predictor of job burnout, a critical consequence of job stress (Hansen, 2012).

Relationship of PsyCap with Employee Engagement and Job Involvement

PsyCap is found to be significantly correlated with employee engagement (Hodges, 2010) and job involvement (Setar, 2013). Optimism and self-efficacy, as dimensions of PsyCap are found to be strongest predictors of employee engagement (Herbert, 2011). PsyCap is found to moderate the impact of work–family–conflict on engagement (Yardley, 2012), suggesting that employees with high PsyCap possess higher levels of employee engagement.

Mediation of PsyCap between Supportive Work Climate and Employee Performance

PsyCap and supportive work climate are necessary for human resources in order to achieve organizational growth. Supportive climate is defined as the total support that an employee receives from his co-workers, other departments and supervisors which helps him with his job demands. Utilizing three diverse samples, Luthans and colleagues showed that employees' psychological capital is positively related to their performance, satisfaction and commitment and a supportive climate is related to employees' satisfaction and commitment (Luthans *et al.*, 2008a). The study's major hypothesis that employees' psychological capital mediates the relationship between supportive climate and their performance was significantly supported.

PsyCap Mediates the Impact of Organizational Justice on Job Satisfaction and Organizational Commitment

According to a recently proposed framework, the perception of higher levels of organizational justice influences PsyCap, which in turn translates into influencing job satisfaction (Totawar, 2013). Drawing support from broaden and build theory (Fredrickson, 1998, 2001), the framework (Figure 7.1) proposes that fairness in distribution of resources triggers positive emotions which build psychological capital (resource). In turn, psychological capital results in increased satisfaction at work.

The framework (Figure 7.1) further contends that the causal effect of organizational justice on commitment is mediated through an employees' PsyCap (Totawar, 2013). Higher the employees' PsyCap higher is the probability that they would remain committed to the organization, despite modest levels of perceived fairness. Four major reasons underlying this proposition are (1) they are hopeful that the fairness situations would improve in the organization; (2) they are optimistic about such a situation and about the overall organizational canvas; (3) they are confident of their own abilities and therefore are less affected by any fairness contingencies in the organization; (4) they are resilient enough to bounce back in case there are any contingencies arising out of unfair situations (Totawar, 2013).

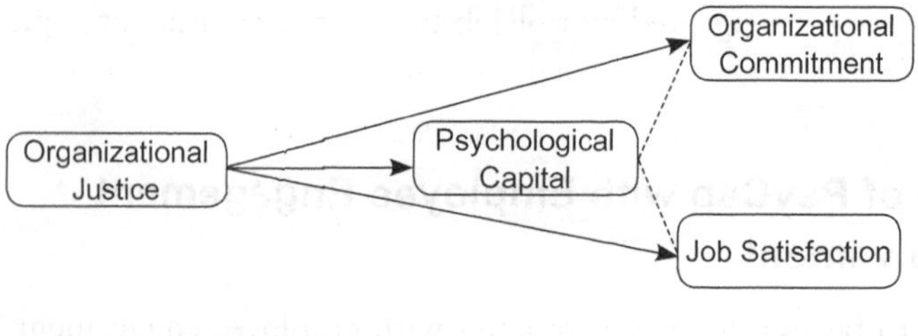

Note: Dotted lines show indirect effects.

Figure 7.1 Mediation of PsyCap. Adapted with permission from '*Effect of organisational justice on job satisfaction and organizational commitment: The mediation of quality of work life and psychological capital*', by Totawar, A. K. (unpublished doctoral thesis), p. 92. Copyright 2013, the Indian Institute of Management, Indore.

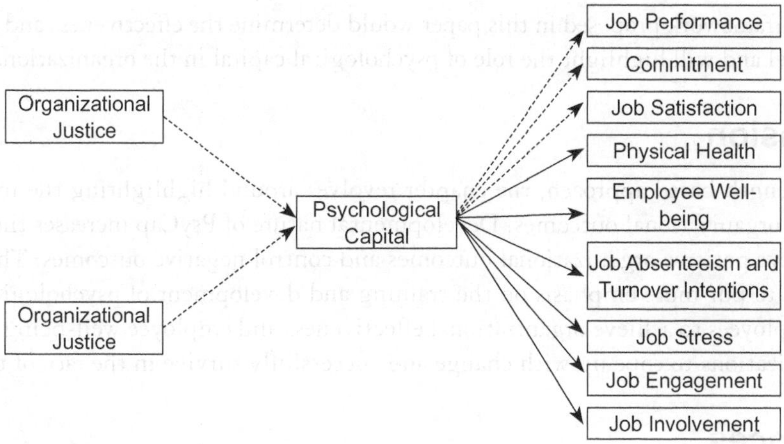

Note: Dotted lines show indirect effects.

Figure 7.2. Antecedents and Consequences of Psychological Capital.

Proposed Conceptual Framework

Based on the review of the literature, the authors propose the following conceptual framework for empirical investigation to examine the antecedents and consequences of psychological capital. The proposed framework is presented in Figure 7.2.

The framework depicts the role of PsyCap as a catalyst to organizations. The model shows two causal paths from work climate and organizational justice on PsyCap representing the antecedents of PsyCap. The model also has an indirect path from PsyCap towards job performance, commitment and job satisfaction. These indirect paths represent the mediating effect of PsyCap between the linkages of work climate–job commitment, organizational justice–commitment and organizational justice–job satisfaction. In addition to the three mentioned indirect paths, the model shows six direct paths from PsyCap towards physical health, employee well-being, job absenteeism and turnover intentions, job stress, employee engagement and job involvement. These paths represent the role of PsyCap as an independent variable on organizational effectiveness and employee well-being. The framework highlights the significance of PsyCap, an individual-based positive capacity in current organizational context.

Implications

The paper suggests that PsyCap is an essential factor for enhancing organizational effectiveness and employee well-being. Considering the developmental nature of PsyCap, HR managers, organizational experts, academicians and policy makers can enhance the PsyCap of employees, students, and employment seekers through teaching, training and development programmes. In an organizational setting, with the help of the proposed framework, HR managers can design organizational HR activities like training and development programmes focusing on development of PsyCap. For academicians and researchers, future empirical research on the theoretical framework proposed in this paper would determine the effectiveness and usefulness of the model and will highlight the role of psychological capital in the organizational context.

Conclusion

Taking strength-based approach, the chapter revolves around highlighting the influence of PsyCap on organizational outcomes. Developmental nature of PsyCap increases the possibility to achieve positive organizational outcomes and control negative outcomes. Therefore, it is essential to put more emphasis on the training and development of psychological capital among employees to achieve organizational effectiveness and employee well-being. This will help organizations to cope up with change and successfully survive in the face of troubles.

References

Abbas, M., and Raja, U. (2011). Impact of Psychological Capital on Innovative Performance and Job Stress. *15th International Business Research Conference.* Retrieved from: http://www.wbiconpro.com/449-Abbas.pdf.

Avey, J. B., Luthans, F., and Jensen, S. M. (2009). Psychological Capital: A Positive Resource for Combating Employee Stress and Turnover. *Human Resource Management*, *48*(5), 677–93.

Avey, J. B., Luthans, F., Smith, R. M., and Palmer, N. F. (2010a). Impact of Positive Psychological Capital on Employee Well-being Over Time. *Journal of Occupational Health Psychology*, *15*(1), 17–28.

Avey, J. B., Luthans, F., and Youssef, C. M. (2010b). The Additive Value of Positive Psychological Capital in Predicting Work Attitudes and Behaviors. *Journal of Management, 36*(2), 430–52.

Avey, J. B., Patera, J. L., and West, B. J. (2006). The Implications of Positive Psychological Capital on Employee Absenteeism. *Journal of Leadership and Organizational Studies*, *13*(2), 42–60.

Avey, J. B., Reichard, R. J., Luthans, F., and Mhatre, K. H. (2011). Meta-analysis of the Impact of Positive Psychological Capital on Employee Attitudes, Behaviors, and Performance. *Human Resource Development Quarterly*, *22*(2), 127–52.

Avey, J. B., Wernsing, T. S., and Luthans, F. (2008). Can Positive Employees Help Positive Organizational Change? Impact of Psychological Capital and Emotions on Relevant Attitudes and Behaviors. *Journal of Applied Behavioral Science*, *44*(1), 48–70.

Avolio, B. J., and Luthans, F. (2006). *The High Impact Leader: Moments Matter in Accelerating Authentic Leadership Development*. New York: McGraw-Hill.

Bandura, A. (1977). Self-efficacy: Toward a Unifying Theory of Behavioral Change. *Psychological Review, 84*(2), 191–215.

Bandura, A. (1982). Self-efficacy Mechanism in Human Agency. *American Psychologist*, *37*(2), 122–47.

Bandura, A. (1989). Human Agency in Social Cognitive Theory. *American Psychologist*, *44*(9), 1175–84.

Bandura, A. (1994). Self-efficacy. In V. S. Ramachaudran (Ed.), *Encyclopedia of Human Behavior* (pp. 71–81). New York: Academic Press.

Bandura, A. (1997). *Self-efficacy: The Exercise of Control*. W. H. Freeman and Company: New York.

Block, J., and Kremen, A. M. (1996). IQ and Ego-resiliency: Conceptual and Empirical Connections and Separateness. *Journal of Personality and Social Psychology*, *70*(2), 349–61.

Bressler, M. (2010). Planning and Projecting Critical Human Resource Needs: The Relationship Between Hope, Optimism, Organizational Commitment, and Turnover Intention among U.S. Army Reserve Soldiers. *Journal of Behavioral Studies in Business, 3*(1), 1–12.

Bycio, P., Hackett, R. D., and Allen, J. S. (1995). Further Assessments of Bass's Conceptualization of Transactional and Transformational Leadership. *Journal of Applied Psychology*, *30*(4), 468–48.

Carver, C. S. and Scheier, M. F. (2002). The Hopeful Optimist. *Psychological Inquiry*, *13*(4), 288–90.

Carver, C. S. (1998). Resilience and Thriving: Issues, Models, and Linkages. *Journal of Social Issues*, *54*(2), 245–66.

Cetin, F. (2011). The Effects of the Organizational Psychological Capital on the Attitudes of Commitment and Satisfaction: A Public Sample in Turkey. *European Journal of Social Sciences*, *21*(3), 373–80.

Chang, C. M., Chen, L. S., Lin, H. Y., and Huang, H. C. (2013). The Relationship between Psychological Capital and Job Satisfaction of University Physical Education Teachers in Taiwan: Job Stress and Job Burnout as Mediators. *Energy Education Science and Technology Part B. Social and Educational Studies*, *5*(3), 1–9.

Cole, K. D., Daly, A., and Mak, A. (2009). Good for the Soul: The Relationship between Work, Well-being, and Psychological Capital. *The Journal of Socio-Economics*, *38*(3), 464–74.

Coutu, D. L. (2002). How Resilience Works. *Harvard Business Review*, *80*(5), 46–50.

Doe, P. J. (1994). Creating a Resilient Organization. *Canadian Business Review*, *21*(2), 22–25.

du Plessis, Y., and Barkhuizen, E. N. (2012). *South African Journal of Economic and Management Sciences*, *15*(1), 16–30.

Dyer, J. G., and McGuinness, T. M. (1996). Resilience: Analysis of the Concept. *Archives of Psychiatric Nursing*, *10*(5), 276–82.

Fredrickson, B. L. (1998). What Good Are Positive Emotions? *Review of General Psychology*, *2*(3), 300–19.

Fredrickson, B. L. (2001). The Role of Positive Emotions in Positive Psychology: The Broaden-and-build Theory of Positive Emotions. *American Psychologist*, *56*(3), 218–26.

Fredrickson, B. L. (2003). The Value of Positive Emotions. *American Scientist*, *91*(4), 330–35.

Hansen, A. A. (2012). Psychological Capital; Subjective Well-being; Burnout and Job Satisfaction Amongst Educators in the Umlazi Region (Masters Thesis). Retrieved from http://researchspace.ukzn.ac.za/xmlui/handle/10413/9418

Harrison, D. A., and Martocchio, J. J. (1998). Time for Absenteeism: A 20 Year Review of Origins, Offshoots and Outcomes. *Journal of Management*, *24*(3), 305–50.

Harrison, D. A., Newman, D. A., and Roth, P. L. (2006). How Important Are Job Attitudes? Meta-Analytic Comparisons of Integrative Behavioral Outcomes and Time Sequences. *Academy of Management Journal*, *4*(2), 305–25.

Herbert, M. (2011). *An Exploration of the Relationships Between Psychological Capital (Hope, Optimism, Self-efficacy, Resilience), Occupational Stress, Burnout and Employee Engagement* (Masters Thesis, University of Stellenbosch). Retrieved from http://scholar.sun.ac.za

Hobfoll, S. E. (2002). Social and Psychological Resources and Adaptation. *Review of General Psychology*, *6*(4), 307–24.

Hodges, T. D. (2010). *An Experimental Study of the Impact of Psychological Capital on Performance, Engagement, and the Contagion Effect* (Doctoral Dissertation). *Retrieved from* http://digitalcommons.unl.edu/businessdiss/7

Home, J. F., and Orr, J. E. (1998). Assessing Behaviors that Create Resilient Organizations. *Employment Relations Today*, *24*(4), 29–39.

Larson, M., and Luthans, F. (2006). Potential Added Value of Psychological Capital in Predicting Work Attitudes. *Journal of Leadership and Organizational Studies*, *13*(2), 75–92.

Liu, L., Chang, Y., Fu, J., Wang, J., and Wang, L. (2012). The Mediating Role of Psychological Capital on the Association Between Occupational Stress and Depressive Symptoms among Chinese Physicians: A Cross-sectional Study. *BMC Public Health*, *12*(219), 1–8.

Luthans, F. (2002a). The Need for and Meaning of Positive Organizational Behavior. *Journal of Organizational Behavior*, *23*(6), 695–706.

Luthans, F. (2002b). Positive Organizational Behaviour: Developing and Managing Psychological Strengths. *Academy of Management Executive, 16*(1), 57–75.

Luthans, F. (2002c). Positive Psychology Approach to OB. In F. Luthans (Ed.), *Organizational behavior* (pp. 286–322). New York: McGraw-Hill/Irwin.

Luthans, F., and Jensen, S. M. (2002). Hope: A New Positive Strength for Human Resource Development. *Human Resource Development Review, 1*(3), 304–22.

Luthans, F., and Youssef, C. M. (2004). Human, Social, and Now Positive Psychological Capital Management: Investing in People for Competitive Advantage. *Organizational Dynamics, 33*(2), 143–160.

Luthans, K. W., and Jensen, S. M. (2005). The Linkage Between Psychological Capital and Commitment to Organizational Mission: a Study of Nurses. *Journal of Nursing Administration, 35*(6), 304–10.

Luthans, F., Avolio, B. J., Walumbwa, F. O., and Li, W. (2005). The Psychological Capital of Chinese Workers: Exploring the Relationship with Performance. *Management and Organization Review, 1*(2), 249–71.

Luthans, F., Avey, J. B., Avolio, B. J., Norman, S. M., and Combs, G. M. (2006). Psychological Capital Development: Toward a Micro-intervention. *Journal of Organizational Behavior, 27*(3), 387–93.

Luthans, F., and Youssef, C. M. (2007). Emerging Positive Organizational Behavior. *Journal of Management, 33*(3), 321–49.

Luthans, F., Youssef, C. M. and Avolio, B. J. (2007). *Psychological Capital: Developing the Human Competitive Edge*. New York: Oxford University Press.

Luthans, F., Youssef, C. M., and Avolio, B. J. (2007). *Psychological Capital*. New York: Oxford University Press

Luthans, F., Avolio, B. J., Avey, J. B., and Norman, S. M. (2007c). Positive Psychological Capital: Measurement and Relationship with Performance and Satisfaction. *Personnel Psychology, 60*(3), 541–72.

Luthans, F., Norman, S. M., Avolio, B. J., and Avey, J. B. (2008a). The Mediating Role of Psychological Capital in the Supportive Organizational Climate–Employee Performance Relationship. *Journal of Organizational Behavior, 29*(2), 219–38.

Luthans, K. W., Lebsack, S. A., and Lebsack, R. R. (2008b). Positivity in Healthcare: Relation of Optimism to Performance. *Journal of Health Organization and Management, 22* (2), 178–88.

Maddux, J. E., and Volkmann, J. (2010). Self-efficacy. In R. H. Hoyle (Ed.), *Handbook of Personality and Self-regulation* (pp. 315–31). Wiley-Blackwell.

Mallak, L. (1998). Putting Organizational Resilience to Work. *Industrial Management, 40*(6), 8.

Masten, A. S. (2001). Ordinary Magic: Resilience Processes in Development. *American Psychologist, 56*(3), 227–38.

Masten, A. S., and Reed, M. J. (2002). Resilience in Development. In C. R. Snyder, and S. J. Lopez (Eds.), *Handbook of Positive Psychology*. New York: Oxford University Press.

Maurer, T. J., and H. R. Pierce. (1998). A Comparison of Likert Scale and Traditional Measures of Self-Efficacy. *Journal of Applied Psychology, 83*(2), 324–29.

Mortazavi, S., Yazdi, S. V. S., and Amini, A. (2012). The Role of the Psychological Capital on Quality of Work Life and Organization Performance. *Interdisciplinary Journal of Contemporary Research in Business, 4*(2), 206–17.

Nath, P., and Pradhan, R. K. (2012). Influence of Positive Affect on Physical Health and Psychological Well-Being: Examining the Mediating Role of Psychological Resilience. *Journal of Health Management, 14*(2), 161–74.

Oi Ling, S. (2013). Psychological Capital, Work Well-being, and Work-life Balance among Chinese Employees: A Cross-lagged Analysis. *Journal of Personnel Psychology, 12*(4), 170–81.

Parker, S. K. (1998). Enhancing Role Breadth Self-efficacy: The Roles of Job Enrichment and other Organizational Interventions. *Journal of Applied Psychology, 83*(6), 835–54.

Peterson, C. (2000). The Future of Optimism. *American psychologist, 55*(1), 44–55.

Raghunathan, R., and Trope, Y. (2002). Walking the Tightrope Between Feeling Good and Being Accurate: Mood as a Resource in Processing Persuasive Messages. *Journal of Personality and Social Psychology*, *83*(3), 510–25.

Roberts, L. M. (2006). Shifting the Lens on Organizational Life: The Added Value of Positive Scholarship. *Academy of Management Review, 31*(2), 292–305.

Scioli, A., Chamberlin, C. M., Samor, C. M., Lapointe, A. B., Campbell, T. L., Macleod, A. R., and McLenon, J. (1997). A Prospective Study of Hope, Optimism, and Health. *Psychological Reports*, *81*(3), 723–733.

Seligman, M. E. P. (1972). Learned Helplessness. *Annual Review of Medicine, 23*, 407–12.

Seligman, M. E. P. (1998). The President's Address. *American Psychologist, 54*, 559–62.

Seligman, M. E. P., and Csikszentmihalyi, M. (2000). Positive Psychology: An Introduction. *American Psychologist, 55*(1), 5–14.

Setar, S. B. (2013). *The Moderating Role of Psychological Capital in the Relationship Between Job Stress and the Outcomes of Incivility and Job Involvement Among Call Centre Employees* (Masters Thesis). Retrieved from http://researchspace.ukzn.ac.za/xmlui/handle/10413/9590

Snyder, C. R. (2000). *Handbook of Hope.* San Diego, CA: Academic Press.

Snyder, C. R. (2002). Hope Theory: Rainbows in the Mind. *Psychological Inquiry*, *13*(4), 249–75.

Snyder, C. R., and Lopez, S. (Ed.). (2002). *Handbook of Positive Psychology.* Oxford, UK: Oxford University Press.

Snyder, C. R., and Lopez, S. J. (2008). *Positive Psychology: The Scientific and Practical Explorations of Human Strengths*. New Delhi: Sage Publications India Ltd.

Snyder, C. R., Feldman, D. B., Taylor, J. D., Schroeder, L. L., Adams, V. H. (2000). The Role of Hopeful Thinking in Preventing Problems and Enhancing Strengths. *Applied and Preventive Psychology, 9*(4), 249–70.

Snyder, C. R., Harris, C., Anderson, J. R., Holleran, S. A., Irving, L. M., Sigmon, S. T., Harney, P. (1991). The Will and the Ways: Development and Validation of an Individual-differences Measure of Hope. *Journal of Personality and Social Psychology*, *60*(4), 570–85.

Snyder, C. R., Sympson, S. C., Ybasco, F. C., Borders, T. F., Babyak, M. A., and Higgins, R. L. (1996). Development and Validation of the State Hope Scale. *Journal of Personality and Social Psychology, 70*(2), 321–35.

Stajkovic, A. D., and Luthans, F. (1998a). Self-efficacy and Work-related Performance: A Meta-analysis. *Psychological Bulletin*, *124*(2), 240–61.

Stajkovic, A. D., and Luthans, F. (1998b). Social Cognitive Theory and Self-Efficacy: Going Beyond Traditional Motivational and Behavioral Approaches. *Organizational Dynamics*, *26*(4), 62–74.

Strumpfer, D. J. W. (2005). Standing on the Shoulders of Giants: Notes on Early Positive Psychology (Psychofortology). *South African Journal of Psychology, 35*(1), 21–45.

Totawar, A. K. (2013). *Effect of Organizational Justice on Job Satisfaction and Organizational Commitment: The Mediation of Quality of Work Life and Psychological Capital* (Unpublished Doctoral Dissertation). Indian Institute of Management Indore, India.

Tugade, M. M., and Fredrickson, B. L. (2004). Resilient Individuals use Positive Emotions to Bounce Back from Negative Emotional Experiences. *Journal of Personality and Social Psychology*, *86*(2), 320–33.

Yardley, K. (2012). *Psychological Capital as a Positive Resource to Assist with the Organizational Outcomes of Work Family Conflict* (Masters Thesis). Retrieved from http://mro.massey.ac.nz/handle/10179/4365

Youssef, C. M., and Luthans, F. 2007. Positive Organizational Behavior in the Workplace: The Impact of Hope, Optimism, and Resilience. *Journal of Management*, *33*(5), 774–800.

CHAPTER 8

Positive Social Support and Flourishing Relationships

Suvashisa Rana and Meena Hariharan

Centre for Health Psychology University of Hyderabad, Telengana

Absract Positive social support is a unique and predominant resource that human being has in order to live life happily in its fullest strength. Supportive and caring family, relatives, friends, neighbours, colleagues, community members, professional groups and even devoted pets are the major pillars of positive social support contributing to the optimal well-being of a person in multifarious ways. Positive social support embellishes positive human functioning resulting in flourishing relationships. These relationships demand certain social obligations and place definite social responsibilities which, though stressful, are considered eustress. At the same time, they also act as catalyst in moderating distress originating from other sources. Positive social support acts as buffer against a cauldron of intense emotions by transforming those into positive life experiences. It is also instrumental in developing positive mindset contributing to the biological, psychological, social and spiritual health of a person. The fabric of eastern society, specifically the Indian traditional social system, offers a number of concrete examples of the contributions of positive social support to individual's psychological resilience and personal growth. In this millennium, where rapid makeover is the basic phenomenon, positive social support nurturing rubric of positive emotions (love, joy, gratitude, serenity, interest, hope, pride, amusement, inspiration, awe), optimism, obligation, forgiveness, altruism and reciprocity offers a charisma for human relationships to flourish.

The professional role of health psychologists that demands regular interaction and experience with hospitalized persons provides good insight into multifarious roles of social support for a person in crisis. The hospitalized persons develop a sense of alienation during the stay in the controlled milieu of hospital, which often create stress, anxiety, loneliness, depression, dependency and pain that is both physical and psychological. Majority of the patients report that the members of their families and significant others share the burden of financial and emotional implications of being sick and hospitalized. Paradoxically, there are also reports from the same that the major source of their stress and interpersonal conflict is the social network. Whether it is a specific situation (hospitalized condition) or normal life course, it is evident that when the basic needs for love, affection and belongingness are not met (Maslow, 1970), one feels lonely, dejected and worthless, all of which causes pain. Perhaps, to counterpoise such feeling human being by nature is gregarious.

Social Support: Meaning

The meaning of 'social support' is simple while the process and functioning of it is complex. Our relationships with others influence our physical, psychological, social and spiritual growth, development and health. This 'relationship with others' is the kernel of social support. The definition of social support is propounded by different branches of psychology, such as developmental, social and health psychology. However, there seems to be a convergence in the connotation of the concept.The term social support refers to the process through which help is provided to others (Feldman and Cohen, 2000). The complexity of the process lies in the fact that it is influenced by characteristics of the social environment and individual participants, transactions that occur between participants, the resources that are provided, and participants' perceptions of these transactions and their implications. The social transaction that is influenced by the above multiple factors may be positive, neutral or negative. Positive social support is defined as those social interactions or relationships that provide individuals with actual assistance or that embed individuals within a social system believed to provide love, or a sense of attachment to a valued social groups or dyad (Rattigan, Walt, Walter and Hobfoll, 2008). Neutral or no social support is a situation where the complexity of influencing factors does not allow any concrete positive outcome. Negative social support refers to a situation where the complex influencing factors results in adding to the crisis or catastrophe and counter production to love, care or affiliation. However, social support is studied as a positive construct rather than negative. Hence, the term 'social support' normally connotes positive social support.

Thus, the term social support refers to supportive relationships that arise between oneself and friends, family members and other persons (Friis and Sellers, 1999). Social support is broadly defined as having one's need to feel a sense of belonging, care, affection and esteem met by others such as spouse, parents, friends, co-workers, and classmates (Durrette and Stern, 2008). This refers to the emotional support. Social support has also been defined as information from others that one is loved and cared for, esteemed and valued, and part of a network of communication and mutual obligations from parents, a spouse or lover, other relatives, friends, social and community contacts, such as churches or clubs (Rietschlin, 1998), or even a devoted pet (Allen, 2003). In this context, social support satisfies the esteem needs. Social support is an interactive process and can be related to altruism, a sense of obligation, and the perception of reciprocity (Schwarzer and Leppin, 1991). When the need for security, affiliation and esteem is gratified, the individual's sense of well-being is enhanced. In the *Encyclopaedia of Positive Psychology*, Gottlieb (2009) defined social support as the social-psychological and interpersonal processes that maintain and promote health and well-being.

The aforesaid definitions of social support clearly distinguish social support from that of social network ties. Friis (2004) bringing a clarity to the two concepts stated that social support denotes perceived emotional support that one receives from, or the value that one attributes to social relationships, whereas the term social network ties is more a quantitative concept that refers the number, structure, or pattern of ties that one has with other people or organizations. The emphasis in case of social support is the 'perception' or 'experience' of the recipient. By stating the difference he postulated that the mere existence of social ties is neither a necessity nor a sufficient condition for social support. However, the above statement does not negate that wider social support network could be a testimony of social competence of the individual and also increases the probability of higher perceived support. While the social network is easily quantified by considering the number of people involved there should also be a method of measuring social support so that it is distinct from the network ties.

Social Support: Approaches

The extent of positive social support can be measured in terms of two approaches (Feldman and Cohen, 2000)—structural approach and functional approach.

Structural support refers to quantitative aspects of relationships, such as the number and interconnectedness of people in a person's social network. That includes objective, reliable, and easy to administer measures (Brissette, Cohen, and Seeman, 2000), such as whether people are married, have relationships with relatives and friends, and belong to social or religious groups. The concept of structural support has received the greatest attention is social integration. This refers to the extent to which persons participate in multiple types of social relationships. Common measures of social integration include the number of social roles that a person occupies and the extent or frequency of social activity that a person is engaged in.

Functional support refers to the resources that other people provide. This could be emotional support (reassurance that we are loved and cared for by others), instrumental or tangible support (assistance with material needs and daily tasks), informational support (receiving guidance and feedback from others) or cognitive support (ideas to solve a problem). Support in any form plays a positive role by contributing to the health and well-being of an individual. However, the prerequisite for this is the parity between the perceived and received social support.

Perceived support is the individual's evaluation of the expected support extended by the members in the social support network in times of need (Heitzmann and Kaplan, 1988). There is substantial evidence that these perceptions are important for health, particularly in times of stress (Cohen and Wills, 1985). Sometimes the perceived support though available may not be adequate; while in other times may not be present. Received support is an interpersonal process that involves the exchange of resources between social network members.

Indicators of received support measure how often different types of support were received over a specific period of time or in the context of a specific event (Heitzmann and Kaplan, 1988). The contribution of social support to health and well-being is positive when perceived support matches the received support, whereas a mismatch between the two fails to bring a desirable impact on health and well-being. Failure to have a positive impact is not as detrimental as having a negative impact. Very often negative support is found to be the consequence of excessive use of the same source of social support (Hariharan and Rath, 2008).

Negative support is sometimes experienced in the face of disadvantage, stress, frustration, conflict and crisis. Such experiences are the consequences of certain intentional negative social interactions also known as social negativity or social conflict. The source of such phenomena are behaviours that are deliberate, negative actions directed toward the individual, which include display of negative affect and negative evaluation of the person, arguments, fights, and blame. Rattigan *et al.*, (2008) called such intentional negative social interactions as 'social undermining'. The explanation of negative social support assumes significance here because sometimes in spite of sustained social relationships, the adversity of the negative support may be so severe as to outweigh the desirable impact of positive support. However, when positive social support is strong and persistent its impact counterbalances, if not overpower the transitory negative support.

Impact of Positive Social Support

Positive social support brings a desirable impact in human life. Literature in the area of social support is very rich in the sense that they identify various desirable outcomes of positive social support. Social support is regarded as resources provided by others (family, relatives, friends, neighbours, colleagues, community members, professional groups and pets), as coping assistance, as an exchange of resources, or even as a personality trait (Schwarzer, Knoll and Rieckmann, 2004). Social support is generally related to better physical and psychological adjustment (Durrette and Stern, 2008). Belonging to a group not only provides psychological benefits, it can improve one's physical health to a good extent. (Hale, Hannum and Espelage, 2005). Expecting support in the future appears to be a stable personality trait that is intertwined with optimism, whereas support provided in the past is based on actual circumstances (Sarason, Levine, Basham and Sarason, 1983). According to Symister and Friend (2003) social support is associated with high self-esteem, which in turn increases optimism and is related to decreased depression. Social support has been found to mediate the relationship between optimism and distress in persons who have experienced a traumatic event (Dougall, Hyman, Hayward, McFeeley and Baum, 2001), as well as in early-stage breast cancer survivors (Trunzo and Pinto, 2003). Studies have shown that social support receipt mediates the relationship between optimism and health (Shen, McCreary and Myers, 2004). Studies have proved the positive impact of social support in the pace of convalescence among cardiac patients. Social support is also found to play a significant role in reducing anxiety in patients who are hospitalized and await angiogram or bypass heart surgery (Chivukula, Swain, Rana and Hariharan, 2013).

Thus, the positive social support acts as a buffer against a cauldron of intense emotions by transforming those into positive life experiences. It is also instrumental in developing positive mindset contributing to the biological, psychological, social and spiritual health of a person.The impact of positive social support can be discussed from two angles. This can be explained through two models, viz., stress-buffering model and main-effect model of social support (Cohen and Wills, 1985).

Stress-buffering Model of Social Support

This model argues that social support acts as prophylactic against life stress. Positive social support acts as a buffer protecting individuals from the harmful effects of stressful conditions. Referring to the stress-moderating effects of social support, Gottlieb (2009) emphasized that the inner spirit of humans are buoyed by interactions with family members and friends, thereby elevating levels of positive affect and self-worth. Viewing social relationships as potentially buffering the individual from life stressors, Cassell (1976) emphasized the importance of physiological processes in mediating the effects of social relationships. The logic that defends this model is that the positive effect that results from close and productive interactions constitutes an effective resource for coping mechanism. When the individual encounters stressful events, one goes into cognitive mediation. Cognitive mediation on the face of stress comprises primary and secondary appraisal of stress. Secondary appraisal is an evaluation process where the individual weighs the personal and social resources under one's disposal. When one finds the social resources adequate, the resultant appraisal of the stress situation is one that of 'challenge' that triggers positive concrete action to mitigate stress. On the other hand, when the social resources are scarce and the internal resources such as ability and skill are absent or inadequate, the resultant appraisal is one that of 'threat' that is accompanied by negative affect and inaction. It is in this context that social support is discussed as a buffering effect because of the perceived functioning of the reserved social resources as a 'buffer'. The impact of this buffer is strong enough to convert stress into eustress or distress.

Social Support, Eustress and Distress

It may not be an exaggeration if one states that positive social support functions as a moderator that has the power to turn stress into eustress. Eustress refers to the positive cognitive response to stress that result in positive affect and a sense of fulfilment. The term was first introduced by Selye (1974). Eustress literally means 'Good Stress' where the prefix 'eu' in Greek refers to good. It is called good stress, because some amount of it is an essential prerequisite to motivate the individual for goal setting achievement behaviour. However, when the degree or level of stress crosses the threshold of its intensity, it turns into distress and disrupts normal functioning.

Coping up with stress is mediated by a cognitive process. The role of positive and negative social support in converting a stress to eustress or distress is explained by the authors here with the help of Figure 8.1.

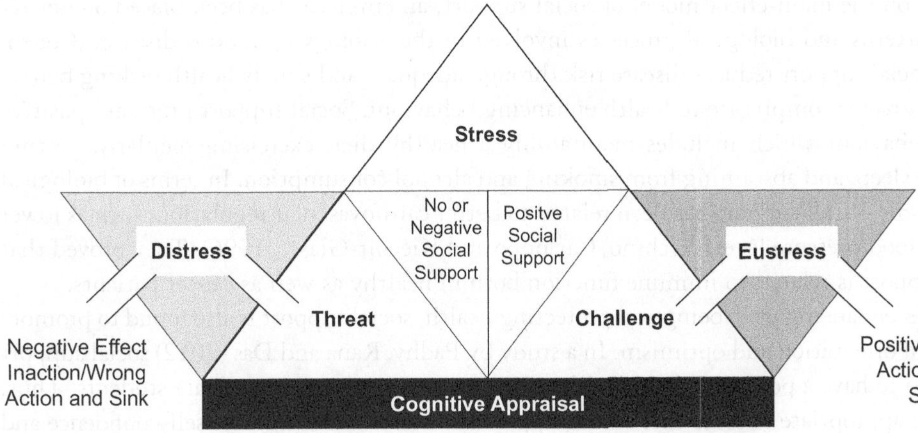

Figure 8.1 Social support, eustress and distress.

The given figure refers to the way the cognitive appraisal of the presence or absence of social support turns life stress into eustress or distress. The figure is an extension of magical triangle propounded by Hariharan and Rath (2008), and Hariharan, Swain and Chivukula (2014). When an individual encounters stress, the stressor goes through a cognitive appraisal. The role of cognitive appraisal is to evaluate the internal (personal abilities, skills, etc.) and external (social support) resources. When the cognitive evaluation identifies no support or negative support, there is perception of 'threat' in the situation causing distress. Distress brings a concomitant negative affect which in turn either inhibits action or triggers ineffective action. As a consequence the probability of succumbing to stress is high. This may end up in 'sinking'. In contrast to this, when the cognitive appraisal identifies external resources in the form of positive social support, the stressor is perceived as a challenge. The perception of challenge has an element of a difficult situation as well as adequate resources to circumvent the same. This leads to the experience of eustress that is associated with positive affect. Positive affect in turn functions as a motivation for action, i.e., active coping behaviour leading to a high probability of effective outcome which helps the individual to sail through the stress successfully.

Main-effect Model of Social Support

Social support exerts direct positive effects on certain physical health outcomes independent of stress (Cohen and Wills, 1985). This hypothesis is known as the main-effect model of social support because it predicts a main effect of social support without a statistical interaction between stress and support. Argument for this model has been obtained when the relationships between structural measures of support (particularly social integration) and health are examined. Studies have found that more socially integrated individuals report greater psychological well-being than their less integrated counterparts (Cohen and Wills, 1985).

Based on the main-effect model of social support, an emphasis has been placed on behavioural patterns and biological processes involved in the etiology of various diseases (Cohen, 1988). Social support reduces disease risk through adequate and timely health seeking behaviour and greater compliance to health enhancing behaviour. Social support promotes positive health behaviour which includes maintaining a healthy diet, exercising regularly, getting adequate sleep, and abstaining from smoking and alcohol consumption. In terms of biological mechanisms, social support has been related to better cardiovascular regulation, such as lower resting blood pressure levels (Uchino, Cacioppo and Kiecolt-Glaser, 1996). They proved that social support is related to immune function both in healthy as well as cancer patients.

Besides enhancing well-being and protecting health, social support is also found to promote hope, goal orientation and optimism. In a study by Padhy, Rana and Das (2012) social support was found to have a positive relation with optimism among the post graduate students. Thus, it may be appropriate to state that social support inculcates and reinforces self-confidence and motivation which are essential requisite for action orientation.

The fabric of eastern society, specifically the traditional Indian social system, offers myriads of concrete examples of the contributions of positive social support to individual's resilience and personal growth. Indian society is highly affiliation-oriented. The positive affect and sense of contentment largely relates to collective action and mutual dependence. Hence, social support contributes positively in creating and sustaining the general positive state. Seeking social support and providing support constitutes a major part of cultural grooming in India. A review of children's literature of Indian origin reflects the cultural significance attributed to social support. The whole series of *Panchatantra* stories for children speak about the role of social support for sustainable well-being. The *Panchatantra* series also provide a good insight on the importance of discretion to be applied in forming flourishing relationships that are mutually aiding. It is very logical to argue that social support has a two-way mutually positive impact where both the provider and the recipient of support initiate and benefit from flourishing relationship.

Flourishing Relationship

Seligman's (2011) well-being theory focuses on five bases of human flourishing—Positive Emotion, Engagement, Relationships, Meaning and Accomplishment (PERMA). 'Relationships' is one of the important pillars of human flourishing, and is both a structural and functional requirement for it. It may be argued that if social support results in the feelings of being cared, loved, esteemed and valued then relationship would flourish with a sense of mutual belongingness, attachment and reciprocity (Cobb, 1976).

In this millennium, where rapid makeover is the basic phenomenon loaded with professional obligation demanding prolonged separations, demolition of joint families, norm of single child, inadequate interpersonal interaction, cut-throat competition leading to breach of trust, betrayal and individualistic achievement orientation replacing affiliation orientation, discussion and debate on flourishing relationship is of great relevance. To flourish means to live within an optimal range of human functioning, one that connotes goodness, generativity, growth and resilience (Keyes, 2002). Ryff and Singer (2003) postulate that conceptions of flourishing articulate the nature of human strengths and well-being and thereby offer enriched formulations of what it means to thrive under challenge. Thus, presence and sustainability of positive relationships with others (social support) is the central point of a flourishing relationship. Ryff and Singer (2000) rightly assumed that interpersonal flourishing is a core feature of quality living across cultures and across time.

To understand how social support contributes to a flourishing relationship, we should focus on the key mechanisms by which social support influences the overall health of a person. Three models, viz., attachment, ecological congruence and conservation of resource, explains the relationship between social support and better health (Rattigan *et al.,* 2008).

Attachment Model

Attachment theory refers to the way that human beings form strong emotional bonds with selected others. This strong connection provides a way to understand personality development, emotional distress and psychopathology. Attachment theorists contend that although attachment characterizes relationships from birth to death, attachment behaviour is often most evident during the early childhood phase of life. Attachment patterns emerge in early childhood and develop in relation to the caregiver. People then adopt these patterns (e.g., being secure, being dependent) in their relations with others in their life. Positive attachments confer positive self-regard, a sense of value, and the sense that others will be there in times of crisis. The lack of positive attachment patterns in childhood creates obstacles to obtaining social support as a resource through the life cycle. Therefore, those who do not form a secure attachment in childhood may be less likely to benefit from or obtain social support as adults. Thus, this model is more or less deterministic.

Ecological Congruence Model

The ecological congruence model proposes that social support interacts with other resources to foster or inhibit stress resistance. While viewing the ecological context as critical in the relation between social support and stress, the specific conditions under which social support exhibits its beneficial and erosive effects become relevant. Specifically, social support influences coping and recovery processes as a function of the individual's characteristics, the context in which the individual resides, the time since the event, and the situational demands. The fit of social support to individuals' needs is partly a function of the unique characteristics of the individual and partly a product of common values shared by individuals of common social systems, such as those found in families, sub-cultures, and shared cultures. In a nutshell, for a resource to be advantageous, it must meet the individual's specific needs and inherent situational demands. Furthermore, the model proposes that a mismatch between the situational demands and cultural values may be counterproductive by aggravating the individual's distress level. Thus, the ecological model is dynamic in nature.

Conservation of Resource Model

Conservation of resource theory has also been used to explain the influence of social support. This theory posits that people are driven to build and maintain resources. Given this innate drive, the perceived or actual loss of resources, or the failure to gain resources after investment, is threatening to them. Four categories of resources are delineated, such as object resources (house, car), personal resources (self-efficacy, self-esteem), energy resources (time invested in studying for a test), and condition resources (e.g., social skillfulness, family relationships, a good marriage). Resources are interdependent, dynamic and can be invested to sustain other resources. This theory suggests that social support is a key resource both because it conveys a sense of attachment and because it provides people with access to resources beyond their personal purview. Social support as a condition resource is additionally important because this act of receiving social support may also build further resources. The theory also suggests that resources and resource demands are interrelated on the individual, familial and community levels. This implies that reciprocity of social support may lead either to building resources beyond the individual or to an additional form of stress and resource depletion. Often, individuals with the greatest need of social support live in communities with those in need of social support and where the emphasis is on reciprocal relationship. In such situations where the individuals are both willing recipients as well as voluntary providers, there is every scope of a flourishing relationship.

Flourishing Relationship: Promotion and Sustenance

Positive social support embellishes positive human functioning resulting in flourishing relationships. According to Snyder and Lopez (2007), 'flourishing relationships are good relationships that continue to get better due to concerted effort of both partners—the provider and receiver'. The path of converting social support to flourishing relationship is gradual and requires understanding of the dynamics of relationships. One needs to be aware of certain common approaches and devote little time from busy schedules to use the existing social support constructively.

Attachment, love and purposeful positive relationship behaviours are the three major components that are useful in creating and sustaining flourishing relationships (Snyder and Lopez 2007). The major prerequisite to create flourishing relationship, as already been delineated is attachment. While the infant-care provider attachment sows the seed for creating flourishing relationship, the attachment in adulthood promotes a sense of security on which a healthy relationship develops. Love is considered as a major expression that promotes and sustains the relationship. Love involves a variety of positive feelings and actions including affection, protection, support and sacrifice (Biswas-Dienerand Dean, 2007). Although the capacity to love and be loved is initially acquired from childhood, it is also a quality that can be developed later in life (Bowlby, 1973). Three purposeful positive relationship behaviours are identified which have to be expressed purposefully to promote flourishing relationships. They are building a mindful relationship connection, creating a culture of appreciation and capitalizing on positive events.

Harvey and Ormarzu (1997) came up with five-component model of *minding relationships*. Minding is the reciprocal knowing process involving non-stop, interrelated thoughts, feelings and behaviours of persons in a relationship. The five components are knowing and being known, making relationship-enhancing attributions for behaviours, accepting and respecting, maintaining reciprocity and continuity in minding. The model explains how intimacy, or the satisfaction and relationship behaviours that contribute to one another's goals in life, may be enhanced. The purposeful positive relationship behaviour of creating a *culture of appreciation* is potentially powerful because of the positive reception of the partner and the partner's behaviour that it promotes and the contemptuous feelings that it prevents (Gottman, 1999). This is based on the 'magic ratio' for marriages which states that five positive interactions to one negative interaction (5:1) are needed to maintain a healthy relationship (Gottman, Murray, Swanson, Tyson, and Swanson, 2003). Creating a culture of appreciation helps to establish an environment where positive interactions and a sense of security are the norms. Expressing positive emotions such as gratitude to a partner is the primary means for creating a positive culture (Snyder and Lopez, 2007). Gable, Reis, Impett and Asher (2004) found that the *process of capitalization* or telling others about positive events in one's life is associated with personal benefits (enhanced positive affect and well-being) as well as interpersonal benefits (relationship, satisfaction and intimacy).

In addition to the above three, there are certain other strategies to build and strengthen rewarding relationships. If these approaches are followed, one would be successful in transforming the inhibition and reaction of social support to pro-action resulting in flourishing relationships. Yeung (2009) explained very clearly the strategies for building rewarding relationships and tackling unsatisfactory relationships. He says:

> 'The key to influencing people successfully is to build strong enough relationships with them that you start to matter to them. If you become an important part of their lives—an ally, a confidant, a friend—then why wouldn't they want to help you out in any way they can?' (p. 106)

According to Yeung (2009), rewarding relationships can be built by responding and reciprocating, understanding the inner most needs, respecting relationship boundaries, building others' esteem, controlling emotional outbursts, facilitating problem-solving and seeing the glass as half full, i.e., always maintaining a positive outlook.

While building healthy and rewarding relationships for promotion of well-being is the initial part of the effort, nurturing and sustaining it is more important so as to prevent a situation where a strain in the relationship itself can be the very source of stress. This is possible by acquiring the necessary social skills and strengthening the social competence. The essential social skills include an acceptable way of broaching disagreement so that it causes no dent in the relationship, the unemotional and rational method of discussing the differences, communicating respect for personality and cultural variations, abiding by the ground rules, restoring the equilibrium in relationships, etc. The underlying principle that guides the above is the unconditional positive regard for others, which prevents adversities in relationships and promotes mutuality in social support leading to flourishing relationships.

References

Allen, K. (2003). Are Pets a Healthy Pleasure? The Influence of Pets on Blood Pressure. *Current Directions, 12*(6), 236–39.

Biswas–Diener, R. and Dean, B. (2007). *Positive Psychology Coaching: Putting the Science of Happiness to Work for Your Clients.* New Jersey: John Wiley & Sons, Inc.

Bowlby, J. (1973). *Attachment and Loss: Vol. 2: Separation: Anxiety and Anger.* New York: Basic Books.

Brissette, I., Cohen, S., and Seeman, T. E. (2000). Measuring Social Integration and Social Networks. In S. Cohen, L. Underwood and B. Gottlieb (Eds.), *Support Measurement and Interventions: A Guide for Health and Social Scientists* (pp. 53–85). New York: Oxford University Press.

Cassell, J. (1976). The Contribution of the Social Environment to Host Resistance. *American Journal of Epidemiology, 104*(2), 107–23.

Chivukula, U., Swain, S., Rana, S., and Hariharan, M. (2013). Perceived Social Support and Type of Cardiac Procedures as Modifiers of Hospital Anxiety and Depression. *Psychological Studies, 58*(3), 242–47.

Cobb, S. (1976). Social Support as a Moderator of Life Stress. *Psychosomatic Medicine, 38*(5), 300–14.

Cohen, S. and Wills, T. A. (1985). Stress, Social Support and the Buffering Hypothesis. *Psychological Bulletin, 98*(2), 310–57.

Cohen, S. (1988). Psychosocial Models of the Role of Social Support in the Etiology of Physical Disease. *Health Psychology, 7*(3), 269–97.

Dougall, A. L., Hyman, K. B., Hayward, M. C., McFeeley, S., and Baum, A. (2001). Optimism and Traumatic Stress: The Importance of Social Support and Coping. *Journal of Applied Social Psychology, 31*(2), 223–45.

Durrette, M. M. and Stern, M. (2008). Cancer Management. In F. T. L. Leong (Ed.), *Encyclopedia of Counseling* (pp. 59–63). California: Sage Publications, Inc.

Feldman, P. J. and Cohen, S. (2000). Social Support. In A. E. Kazdin (Ed.), *Encyclopedia of Psychology: Vol. VII* (pp. 373–76). New York: Oxford University Press.

Friis, R. H. (2004). Social Support. In W. E. Craighead and C. B. Nemeroff (Eds.), *The Concise Corsini Encyclopedia of Psychology and Behavioral Science* (3rd Ed., pp. 919–21). New Jersey: John Wiley and Sons, Inc.

Friis, R. H., and Sellers, T. A. (1999). *Epidemiology for Public Health Practice (2nd Ed.).* Gaithersburg, MD: Aspen.

Gable, S. L., Reis, H. T., Impett, E. A., and Asher, E. R. (2004). What Do You Do When Things Go Right? The Intrapersonal and Interpersonal Benefits of Sharing Positive Events. *Journal of Personality and Social Psychology, 87*(2), 228–45.

Gottlieb, B. H. (2009). Social Support. In S. J. Lopez (Ed.), *The Encyclopedia of Positive Psychology* (pp. 913–16), West Sussex: United Kingdom Blackwell Publishing Ltd.

Gottman, J. M. (1999). *The Seven Principles for Making Marriage Work.* New York: Crown.

Gottman, J. M., Murray, J. D., Swanson, C., Tyson, R., and Swanson, K. R. (2003). *The Mathematics of Marriage: Dynamic Nonlinear Models.* Cambridge: MIT Press.

Hale, C. J., Hannum, J. W., and Espelage, D. L. (2005). Social Support and Physical Health: The Importance of Belonging. *Journal of American College Health, 53*(6), 276–84.

Hariharan, M. and Rath, R. (2008). *Coping with Life Stress: The Indian Experience.* New Delhi: Sage Publications India Pvt. Ltd.

Hariharan, M., Swain, S., and Chivukula, U. (2014). Childhood Stress and Its Impact on Learning and Academic Performance. In A. J. Holliman (Ed.), *Educational Psychology* (pp. 127–39). London: Routledge.

Harvey, J. H., and Ormarzu, J. (1997). Minding the Close Relationship. *Personality and Social Psychology Review, 1*(3), 224–40.

Heitzmann, C. A. and Kaplan, R. M. (1988). Assessment of Methods for Measuring Social Support. *Health Psychology,* 7(1), 75–109.

Keyes, C. L. M. (2002). The Mental Health Continuum: From Languishing to Flourishing in Life. *Journal of Health and Social Behavior, 43*(2), 207–22.

Maslow, A. (1970). *Motivation and Personality.* New York: Harper and Row.

Padhy, M., Rana, S., and Das, S. (2012). Social Support and Optimism Among Students. Indian *Journal of Psychometry and Education, 43*(1), 34–38.

Rattigan, S. S., Walt, L. C., Walter, K. H., and Hobfoll, S. E. (2008). Cancer Management. In F. T. L. Leong (Ed.), *Encyclopedia of Counseling* (pp. 872–74). California: Sage Publications, Inc.

Rietschlin, J. (1998). Voluntary Association Membership and Psychological Distress. *Journal of Health and Social Behaviour, 39*(4), 348–55.

Ryff, C. D. and Singer, B. (2000). Interpersonal Flourishing: A Positive Health Agenda for the New Millennium. *Personality and Social Psychology Review, 4*(1), 30–44.

Ryff, C. D. and Singer, B. (2003). Flourishing under Fire: Resilience as a Prototype of Challenged Thriving. In Corey L. M. Keyes and Jonathan Haidt (Eds.), *Flourishing: Positive Psychology and the Life Well-lived* (pp. 15–36). Washington, DC, US: American Psychological Association.

Sarason, I., Levine, H. M., Basham, R. B., and Sarason, B. (1983). Assessing Social Support: The Social Support Questionnaire. *Journal of Personality and Social Psychology, 44*(1), 127–39.

Schwarzer, R., and Leppin, A. (1991). Social Support and Health: A Theoretical and Empirical Overview. *Journal of Personal and Social Relationships, 8*(1), 99–27.

Schwarzer, R., Knoll, N., and Rieckmann, N. (2004). Social Support. In A. Kaptein and J. Weinman (Eds.), *Health Psychology* (pp. 158–82). Oxford, England: Blackwell.

Seligman, M. E. P. (2011). *Flourish.* New York: Simon and Schuster, Inc.

Selye, H. (1974). *Stress Without Distress.* Philadelphia: J. B. Lippincott Co.

Shen, B. J., McCreary, C. P., and Myers, H. F. (2004). Independent and Mediated Contributions of Personality, Coping, Social Support, and Depressive Symptoms to Physical Functioning Outcome Among Patients in Cardiac Rehabilitation. *Journal of Behavioral Medicine, 27*(1), 39–62.

Snyder, C. R. and Lopez, S. J. (2007). *Positive Psychology: The Scientific and Practical Explorations of Human Strengths.* New Delhi: Sage Publications India Pvt. Ltd.

Symister, P. and Friend, R. (2003). The Influence of Social Support and Problematic Support on Optimism and Depression in Chronic Illness: A Prospective Study Evaluating Self-esteem as a Mediator. *Health Psychology, 22*(2), 123–29.

Trunzo, J. J., and Pinto, B. M. (2003). Social Support as a Mediator of Optimism and Distress in Breast Cancer Survivors. *Journal of Consulting and Clinical Psychology, 71*(4), 805–11.

Uchino. B. N., Cacioppo. J. T., and Kiecolt-Glaser, J. K. (I996). The Relationship Between Social Support and Physiological Processes: A Review with Emphasis on Underlying Mechanisms and Implications for Health. *Psychological Bulletin, 119*(3), 488–31.

Yeung, R. (2009). *Emotional Intelligence: The New Rules.* London: Marshall Cavendish Limited.

CHAPTER 9

Positive Work–Family Interface

N. K. Chadha and Vandana Gambhir nee Chopra

Department of Psychology, University of Delhi, Delhi

Abstract Work and family are the two dominant inextricably intertwined spheres of an individual's life. Positive commitment and balanced involvement in both work and non-work responsibilities are essential to live a happy, satisfying and productive life. On the other hand, conflict between the two is detrimental for individual, family, organizations and society. The present chapter aims to explore positive bi-directional work-to-family enrichment (WFE) and family-to-work enrichment (FWE) through the lens of positive psychology. Closely looking at the structure and organization of work and family, we will examine the changing relations and potential sources of spillover and conflict between the two. Special attention is given to work–family relationships with conceptions of gender. Seeking a more balanced approach to the positive work–family enrichment, we articulate conditions that foster positive work–family interface and reveal how the interface helps an individual to craft triumphs across multiple domains of life.

Work and family are two most significant domains of an individual life that share interconnecting and reciprocal influences on each other. Much has been written in the scholarly texts about the relationship between work and family—the two prominent provinces of a person's life (Eby, Casper, Lockwood, Bordeaux and Brinley, 2005). Family is an integral element of an individual's life that teaches the value of love, affection, care, truthfulness and self-confidence and provides tools and suggestions necessary to succeed at work. On the other hand, the ability for people to do work has a multitude of benefits for the family including, for example, an opportunity to earn one's livelihood, to contribute positively and meaningfully to the society, and to satisfy one's own drive for achievement. Family and work-related outcomes together make positive impact on people's outlook towards life and overall health status.

Early researchers treated work and family systems as if they operated independently (Parsons and Bales, 1955). Men traditionally assumed the role of the breadwinner and women that of the homemaker. In the past few years, due to changing social scenario, work and family domains have merged to an extent that it is difficult to consider any one of them in isolation to the other. Changing composition of the workforce with greater involvement of women in the active working force (Lerner, 1994), more number of working single-parent, dual-earner families (Bumpass, 1990) have made work and family responsibilities and commitment to stand in unison to each other. Today, the two domains reflect an individual's orientation across different life roles, an inter-role phenomenon (Marks and MacDermid, 1996). An individual 'has' and 'should' demonstrate equally positive commitments to both the roles and should hold a balanced orientation towards both with an attitude of attentiveness and care. Thus, positive work–family interface has important implications for individuals, organizations, and society, and consequently, a growing body of research has explored the intersection of work and family domains.

The aim of the present chapter is to review the positive side of work–family interface. Several researchers have called for a balanced approach to the work–family interface by examining the benefits of multiple role memberships (Frone, 2003; Parasuraman and Greenhaus, 2002). They argued that positive commitment and balanced involvement in both work and non-work responsibilities are essential to live a happy, satisfying and productive life. In contrast, work–family conflicts are regarded as time strains, missed work or family activities, and the spillover of stress from work to home or vice versa. This chapter examines both, changing relations and potential sources of spillover and conflict and positive interface and enrichment between work and family domains. A special section of the chapter is devoted to work–family relationships with conceptions of gender. Seeking a more balanced-approach to the positive work family enrichment, we articulate conditions that foster positive work family interface and reveal how the interface helps an individual to craft triumphs across multiple domains of life.

Work and Family: A Bi-directional Interface

Researchers confirm that there exists a two-way relationship between work and family provinces, both influencing each other in positive and negative ways (Greenhaus and Beutell, 1985; Kinnunen and Mauno, 1998). Work and family responsibilities meddle with each other's work roles and interfere in fulfilling family duties and family responsibilities with work life. This leads to what has been defined in literature as work–family conflict. Researches correlating workplace variables with family processes have found that marital satisfaction is greatly affected by work pressure and family pressure hampers job performance and job satisfaction (Crouter, 1984; Frone, Russell, and Cooper, 1992; MacEwen and Barling, 1994; Parasuraman, Purohit, Godshalk, and Beutell, 1996).

However, the relationship of work and family is not always negative. A parallel body of research suggests that there exists a synergistic and harmonious relationship in which both domains complement each other, family balancing out the stress and strains of work life and work giving support to live a healthy, happy life. Barnett (1996) and Sieber (1974), suggested that performing work and family roles simultaneously offers a wide arena of resources and prospects to a person that enhances his growth and leads to optimal functioning. A healthy marital life has been considered an important safeguard for job related stress, especially for men (Adams, King, and King, 1996; Barnett, Marshall, and Pleck, 1992; Barnett, Marshall, Raudenbush, and Brennan, 1993; Gattiker and Larwood, 1990; O'Neil and Greenberger, 1994; Weiss, 1990).

Recent literature on work/non-work interface has moved away from classical distinction between the two domains. Rather, it has been found that clear distinction between the two domains is not possible conceptually and empirically. Both domains function simultaneously and positive and negative interface between the two depends on workplace attributes, family situations and personal disposition or individual differences. Work–family interface consists of both aspects, conflict as well as synergy depending on whether the two significant domains of an individual's life are in contradiction or in harmony to each other.

Work–Family: The Conflict Orientation

The conflict orientation of work–family literature holds the view that the two domains are irreconcilable and contrary to each other. Work–family conflict results because the demand of one domain clashes with the demand of the other and a person is not able to restore a balanced approach between the two. As a result of incongruity, a person experiences stress, time pressure, neglect of work and family activities and transfer of tension from work to home or vice-versa.

The Role Strain Hypothesis of Work–family Conflict: The work/non-work interface has been classically studied in literature from *role strain hypothesis*. According to this perspective, work–family conflict is 'a form of inter-role conflict in which role pressures from the work and family domains are mutually incompatible in some respect' (Greenhaus and Beutell, 1985, p. 77). This theory presumes that an individual engaged in multiple roles in family and workplace settings experiences stress and strain due to inter-role conflict. According to Frone, Russel and Cooper (1997), the two spheres are unable to get along because the person is unable to manage different role characteristics as involvement in one role (either family or work) makes difficult to get involved in the other role.

Conflict between Work and Family (WFC) Is Distinct from Conflict between Family and Work (FWC)

Work–family conflict was considered to be a unidimensional construct in earlier research findings. Empirical studies during 1980s documented how work-related variables impact family life. However, in recent years, several studies have advanced the understanding of how work affects family life and vice-versa. It was proposed that the domains share the reciprocal relationship with each other and conflict can arise in either domain affecting the other. A conceptual distinction between work conflicting with the family and family conflicting with work has been made. The former has been defined as work to family conflict (WFC) and the latter as family to work conflict (FWC). A person doing overtime, spending 14–16 hours in job will experience work to family conflict and a person unable to attend office due to family obligations will face family to work conflict. In such processes of role interaction between work and family, attitudes and behaviour carry over from one another, often provoking competing (and conflicting) demands resulting in stress (Greenhaus and Beutell, 1985). Empirical reports from different samples indicate that work to family conflict and family to work conflict are at best, moderately correlated (r = .30–.55; Frone, Yardley and Markel, 1997; Gutek, Searle and Klepa, 1991).

WFC and FWC, the related forms of inter-role conflict are based on *the scarcity perspective* of Marks (1977). According to this perspective, involvement and fulfilment of multiple roles of work and non-work requires time and energy resources of a person. Since there is limited availability of time and resources, engagement in multiple roles is likely to be associated with exhaustion of these scant resources leading to conflict. A balanced allocation of time and resources to work and non-work domains can help a person to better utilize the limited resources and avoid role conflict.

Sources of WFC and FWC

Although several sources of WFC and FWC have been identified, most researchers agree that the time devoted to a given role, the strain produced by a given role and the general demands of a role are domain elements of WFC and FWC (Bachrach, Bamberger and Conley, 1991; Greenhaus, 1988; Greenhaus and Beutell, 1985; Voydanoff, 1988). Three sources of conflict are as follows:

Time-based Conflict

This type of conflict results due to time pressures from one domain which makes it impossible for a person to devote time to another domain. Time-based conflict occurs when the amount of time devoted to the work (or family) role interferes with performing family (or work)-related responsibilities. A person experiencing time-based conflict gets preoccupied either in work role (like, time-based submission of work) or in family role (participation in family activities) to an extent that the functioning of the other domain is affected.

Strain-based Conflict

This type of conflict results due to over engagement in one role leading to stress, strain, anxiety, tension, irritability or depression. Strain experienced by undue involvement in one role makes it difficult for a person to meet the roles and demands of other roles leading to conflict. For example, irritability and anxiety created by work interfere with performing family duties and vice versa.

Role-based Conflict

The general demands of a role refer to the responsibilities, requirements, expectations, duties, and commitments associated with a given role. Role-based conflict results because of display of specific type of behaviour in a role which does not meet expectations regarding behaviour in that role. A person exhibiting casual, open and personal attitude at workplace with colleagues might experience a behaviour-based conflict since this is not acceptable according to the office norms. Similarly, a person keeping strict, logical and work-related attitude at home will experience behaviour-based conflict.

Depending upon the etiological studies of conflict between work/non-work domains, WFC has been characterized as a form of inter-role conflict in which the general demands of time devoted to, and strain created by the job interfere with performing family-related responsibilities. FWC is a form of inter-role conflict in which the general demands of, time devoted to, and strain created by the family interfere with performing work-related responsibilities.

Impact of WFC and FWC on Work and Family Related Outcomes

WFC and FWC have been shown to have a negative impact on important work and family-related outcomes that in turn influence general health and well-being. Researchers have examined relationships of WFC and FWC with various on- and off-job variables

Burke (1988) conducted a study on a sample of police officers to find out how WFC is related to on-job variables. The study reported that higher levels of work interfering with family leads to psychological burnout, alienation and low job satisfaction. Similarly, Bacharach, Bamberger and Conley (1991) conducted their study on nurses and on a sample of engineers and found that work interfering with family was significantly related to burnout, which then was related to lower job satisfaction. Thomas and Ganster (1995) reported that WFC was negatively related to job satisfaction and positively related to depression and health complaints among health care workers.

Empirical researches report that work interfering with family had a significant relationship with family-related variables. Family-related effects were extensively studied by Higgins and his colleagues (Duxbury and Higgins, 1991; Higgins and Duxbury, 1992; Higgins, Duxbury, and Irving, 1992) by carrying out study on dual-career families. In a study of 220 career-oriented individuals, Higgins, Duxbury and Irving (1992) found that work interfering with family resulted in poor quality of family life. This poor quality of family life resulted in poor levels of life satisfaction among workers.

Several empirical studies have also supported relationship between family interfering with work. For instance, Wiley (1987) noted that family interfering with work was negatively related to job satisfaction, organizational commitment, and life satisfaction in a sample of employed graduate students. Along these same lines, Frone, Russell and Cooper (1992) documented that a higher level of family interfering with work was related to incidences of clinical depression and distress on the job for a large, community-based sample of working adults.

In summary, the literature generally suggests that both WFC and FWC have negative impact on employee health factors as well as workplace productivity. In particular, concerns include increased family tension, increased stress of both the employee and other members of the family, increased occurrence of stress-related illness and absence, and increased use of medical care resources including medications, behavioural health services, and clinical tests for psychosomatic illnesses. Clearly, WFC and FWC have a relationship with both domain-specific work related and general measures of health and well-being.

Work–family: A Synergistic Viewpoint

While it is clear that in some circumstances pressure from one's life setting (e.g., work or family) may spillover and undermine functioning in another. A parallel body of studies also confirms a harmonious and facilitating relationship between the two. Considerable evidence within the work–family literature suggests that the work–family interface can also result in synergies, or 'positive spillover' between work and family (Barnett, Marshall and Pleck, 1992; Pope, 1991; Seiber, 1974). Scholars typically conclude that having the opportunity to talk through difficulties at work, or having a partner who is sensitive to job-related pressures may help individuals better handle the pressures associated with their jobs and consequently perform better (Barnett, 1996; Gattiker and Larwood, 1990; Weiss, 1990). On the other hand, development of decision-making skills at work may help a person to use that art at home in effectively handling the issues related to family and children (Crouter, 1984). Gender studies on work–family interface have empirically proved that employed, married mothers experience greater physical health and happiness when compared to unemployed, married mothers (Waldron, Weiss and Hughes, 1998).

The Role Enhancement Hypothesis of Positive Work–family Interface

A number of concepts have emerged to explain the positive relation between work and family provinces of life. These concepts are based on *the role enhancement hypothesis* which suggests that involvement in multiple different roles offers a variety of opportunities and resources to a person to utilize his skills and energy for personal growth and better functioning. The positive side of work/non-work research has shifted its focus from role-conflict to role-enhancement or facilitation. As opposed to *scarcity perspective* of role strain hypothesis, role enhancement hypothesis is based on Marks's *expansion approach* (Marks, 1977). *The expansion approach* believes that performing and fulfilling expectations of multiple roles is not always associated with role strain and depletion of human energy and resources. Laying his assumptions from the field of human physiology, Marks reported that efficiently managing multiple roles may also generate more energy as the process of energy depletion simultaneously works with the process of energy production. A person adequately performing multiple roles simultaneously initiates the process of energy production along with energy expenditure which helps him to perform diverse roles efficiently. With regard to resources, it was suggested that each role offers a set of resources that helps an individual to fulfil demands associated with other roles. Thus, employment is a resource to fulfil demands of family and, in turn, family is a source of psycho social support to carry out job proficiently.

Terminology of Positive Work Family Interface

The positive synergies between work and family have been reported under a variety of different labels (Greenhaus and Powell, 2006; Grzywacz and Butler, 2005) in literature. Some of these constructs of positive work family interface include enrichment, positive spillover, work–family enhancement, work–family fit, work–family integration, work–family balance and facilitation. Though defined differently, all concepts measure the positive side of work–family interface at the individual level of analysis.

Work–Family Balance

The most widely used and contested term in the literature is work–family balance. Work–family balance refers to the extent to which an individual is equally engaged in—and equally satisfied with—his or her work role and family role (Greenhaus, Collins and Shaw, 2003). These researchers identified three key components inherent in this balance: time balance (i.e., equal time devoted to work and family roles), involvement balance (i.e., an equal level of psychological involvement in work and family roles) and satisfaction balance (i.e., an equal level of satisfaction with work and family roles).

Work–Family Enrichment

Another popular concept of positive work–life interface is Greenhaus and Powell's (2006) 'work–family enrichment', which is defined as the extent to which the quality of life is improved by experiences in the other role. Greenhaus and Powell reviewed 19 studies that measured the positive side of the work–family interface and assessed the positive effect of experiences in one role on experiences or outcomes in another role (p. 74).

Similar to work–family conflict, the concept of work–family enrichment is also considered as bidirectional in nature (Frone 2003). That is, benefits can be derived from work and applied to family {work-to-family enrichment (WFE)} or derived from family and applied to work {family-to-work enrichment (FWE)}. According to Greenhaus and Powell's (2006) model, diverse role experiences in both domains offer five categories of resources that operate directly or indirectly producing enrichment. These five resources are—*skills and perspectives* (e.g., interpersonal skills, coping skills, respecting individual differences), *psychological and physical resources* (e.g., self-efficacy, hardiness, optimism), *social-capital resources* (e.g., networking, information), *flexibility* (e.g., flexible work arrangements) and *material resources* (e.g., money, gifts). Greenhaus and Powell's theoretical model helps elucidate work–family enrichment process by means of resources and positive affect. Resources generated in one domain can either have a direct instrumental effect or indirect effect on performance in another domain through positive affect. That is, positive affect generated by happy experiences within the family may produce more positive affect in work role and vice-versa.

Work–Family Fit

The concept of *'work–family fit'* is based on *Barnett's fit model* which emphasizes the importance of *fitness* as an intervening process in the relationship between the work domain and family domain. 'Fit' is conceptualized as, 'the extent to which workers realize various components of their work–family strategies, that is, their plans for optimizing their own work and non-work needs as well as those of other members of their work–family/social system' (Barnett, Gareis and Brennan, 1999, p. 307). The *fit* construct does not assume an inherent conflict between family and work domains. Instead, it believes in adaptive strategies that simultaneously maximize employees' ability to meet the needs of the workplace and their ability to meet the demands of family system.

Impact of Work–Family Enrichment (WFE) and Family–Work Enrichment (FWE) on Work and Family Related Outcomes

Likewise, work–family conflict, researchers have proposed a number of different consequences of WFE and FWE. Empirically, work–family synergy is shown to have significant outcomes on work-related, non-work-related and health-related variables.

1. Enrichment has been related to a number of important *work related outcomes*, including job satisfaction (i.e., degree of pleasure derived from the job) and affective commitment (i.e., emotional attachment to the organization). Based on Greenhaus and Powell's (2006) affective path, if a parent perceives resources stemming from his or her family role (e.g., better time management skills as a result of parenting), he or she is likely to be a better parent, which in turn creates more positive emotions at home, and this translates to more positive emotions at work (i.e., FWE). Conversely, resources acquired at work (e.g., self-esteem) may result in better performance at work, which has the effect of creating more positive affect at work, ultimately transferring to more positive affect in the family domain (i.e., WFE). Thus, in both cases, individuals who experience joy in their work have higher job satisfaction and affective commitment.
2. Positive effects of work family synergy on *non work-related outcomes* have been reported in several empirical studies. Enrichment has positive effects on one's life outside of work as well as the aforementioned benefits in the work domain. Enrichment is known to be positively related to both family and life satisfaction (e.g., van Steenbergen, Ellemers, and Mooijaart, 2007). Involvement with family domain activities has been linked to greater life satisfaction (Judge, Boudream and Bretz, 1994). According to Greenhaus and Powell (2006), family participation is likely to generate a variety of resources. For instance, coping skills generated in a family role should strengthen one's

performance and positive affect at home, and in turn improve one's positive affect at work. It follows that greater positive feelings and emotions about a family role should result in greater family happiness and contentment in life.

3. Williams, Franche, Ibrahim, Mustard and Layton (2006) found that greater enrichment was related to *better physical health*, perhaps because these individuals have a 'solid resource reservoir' that makes them better equipped to handle stress, which leads to greater well-being. Research has suggested that participation in multiple roles can buffer the negative effects of one role on the other (e.g., Sieber, 1974). Moreover, enrichment can generate resources that may be essential for coping with stress-related variables (Greenhaus and Powell, 2006). Summarizing, it can be said that work–life synergy is related to important personal and organizational outcomes. Participation in one role may enrich the quality of life in a second role. In particular, individuals who perceive greater enrichment seem to reciprocate with more favourable attitudes towards the originating role. Thus, efforts should be made not only to reduce perceptions of work–family conflict but also to increase perceptions of work–family balance.

Gender with Respect to Work/Non-work Interface

Issues of gender have been a central focus of work–family scholarship during the past three decades. Many studies indicate that men and women report similar levels of work–family conflict (Barnett and Gareis, 2006) but seem to exhibit different behaviour patterns in response to this conflict (Mennino and Brayfield, 2002). Studies have identified gender differences in work–family stress, showing significantly more conflict for women than for men (Duxbury and Higgins, 1991). For example, when children are young, men are comfortable and adaptable to changes in timing and location of work, whereas women may need shorter working hours or part-time job options. Further, expectations of mothers and fathers continue to differ, even for employed parents. Employed mothers may experience greater work–family conflict, given the continued gender differences in time spent in care giving and household labour. Besides in the Indian scenario women are expected to be more dedicated to their families, devoting much of their time and focus to the needs of the families. Some research reports that fathers have too little time for children—a specific measure of time inadequacy—due to gender differences in the amount of time spent in paid work and away from children. However, consistent with intensive mothering expectations, mothers' well-being suffers more than fathers' when they feel they do not have enough time with children or a spouse.

Work–family synergy also has a differential impact on various outcomes depending on gender differences. As a result of gender role socialization, there is evidence to suggest that men and women may view their work and family roles differently. For example, women are more likely to integrate work and family roles whereas men are more likely to segment or mentally separate these roles (Andrews and Bailyn, 1993). Women may be more likely to adopt this integrative model in order to balance work and family demands. Even though gender roles have shifted, women continue to spend more time on household activities and childcare, and make more adjustments to their work schedules than men (Friedman and Greenhaus, 2000). Wayne, Randel and Stevens (2006) model of work–family facilitation notes that men and women use resources differently, and this may also be a product of gender socialization.

Thus, gender constitutes the socio-demographic characteristic that has been most frequently examined with respect to the prevalence of the various dimensions of the work/non-work interface. The reason is the generally believed gender role expectation that work is more important for men and family life is more important for women. Because of higher family responsibilities, family factors intrude into the work situation more often in the case of women than in the case of men (home negatively influencing work), whereas work factors intrude into the family situation more often in men's than in women's case (work negatively influencing home).

Strategies to Improve Positive Work Family Interface

Due to increasing number of employees experiencing work–family conflicts, more focus is now diverted towards strategies that can foster integration between work and family lives. The three most common strategies in this regard are: (1) more flexible working time patterns; (2) facilitation of leave arrangements and (3) provision of childcare facilities. Factors that have been identified as triggering factors that motivate companies towards action are the percentage of female workers, the tightness of the labour market, and, most importantly, the extent to which companies wish to create a committed workforce. Some relevant organizational and personal strategies helpful to maintain work family synergy are reported below.

Organizational Strategies

It's not easy to identify interventions or changes in company policy that guarantee improvement in the work–family balance profile of workers. For example, a reduction in the number of hours worked may not necessarily improve an employee's work–family balance or reduce work–family conflict. Furthermore, two employees working the same number of hours may experience very different levels of conflict. However, if the company allows supervisors to work with individual employees to optimize the work schedule to fit with family needs, major improvements may be achieved in the perceptions of conflict. A number of approaches reported in literature that can be used to address the challenge of work–family balance are: flexible scheduling of working hours, leave policies, childcare, family-friendly organizational culture, supportive supervisor, supportive co-workers, tax/benefit policies, job sharing and telecommuting.

A study from Cohen (1997) among 300 employees of a school district in Canada showed that organizational commitment of employees (particularly of those who find their non-work domains important) was particularly dependent on how organizations react to the non-work domains of employees (e.g., considering employees' personal lives when making important decisions about careers, accommodating employees' special non-work needs, having a flexible attitude with respect to employees' work schedules). Cohen's findings imply that organizations may increase positive attitudes among their employees by showing more respect for their non-work domains.

Personal Strategies

Strategies that were aimed at positive attitudinal changes in self with respect to what demands can realistically be met in both domains seemed to be more effective in coping with high demands from both domains and improving family synergy than strategies aimed at changing the attitudes or behaviours of others. Beutell and Greenhaus (1983) have studied the effectiveness of three types of coping strategies for dealing with work/non-work conflict among 115 married women (with at least one child) who were attending college. Their study showed that active attempts to change the structural and/or personal definition of one's roles were more effective in dealing with work/non-work conflict than more passive and reactive role behaviour. They describe the active reaction pattern as attempts to lessen the conflict by discussing and finding mutual agreement with other people about what can be expected from them, as well as redefining one's own attitudes and perceptions in a more realistic way. The passive reaction pattern can be described as relying on existing role behaviours and trying to meet the unchanged expectations of others as well as of oneself.

Conclusion

Work and family are interwoven. Work family interface is a multifaceted phenomenon with work and family mutually influencing each other in both negative as well as positive ways. Though difficult to conceptualize, work–family interface is an important topic to study, not only because an increasing number of workers have difficulty in balancing work and non-work demands, but also, because the effects of work are not restricted to the workplace and the effects of the non-work domain affect employees' behaviour and experiences at work. However, it is vital to expand our understanding of the phenomenon by considering how work positively affects the non-work domain and how non-work domain (i.e., family) can facilitate functioning at work. This can be seen also as an expression of a more general trend towards positive psychology that focuses on human strengths and optimal functioning rather than on weaknesses and malfunctioning.

References

Adams, G. A., King, L. A., and King, D. W. (1996). Relationships of Job and Family Involvement, Family Social Support, and Work–family Conflict with Job and Life Satisfaction. *Journal of Applied Psychology, 81*(4), 411–20.

Andrews, A., and Bailyn, L. (1993). Segmentation and Synergy: Two Models Linking Work and Family. In J. C. Hood (Ed.), *Men, Work, and Family* (pp. 262–75). Newbury Park, CA: Sage.

Bacharach, S. B., Bamberger, P., and Conley, S. (1991). Work–Home Conflict among Nurses and Engineers: Mediating the Impact of Role Stress on Burnout and Satisfaction at Work. *Journal of Organizational Behavior, 12*(1), 39–53.

Barnett, R. C. (1996). *Toward a Review of the Work/Family Literature: Work in Progress.* Boston, MA: Wellesley College Center for Research on Women.

Barnett, R. C., Marshall, N. L., and Pleck, J. H. (1992). Men's Multiple Roles and Their Relationship to Men's Psychological Distress. *Journal of Marriage and the Family, 54*(2), 358–67.

Barnett, R. C., Marshall, N. L., Raudenbush, S. W., and Brennan, R. T. (1993). Gender and the Relationship Between Job Experiences and Psychological Distress: A Study of Dual-earner Couples. *Journal of Personality and Social Psychology, 64*(5), 794–806.

Barnett, R. C., Gareis, K. C., and Brennan, R. T. (1999). Fit as a Mediator of the Relationship Between Work Hours and Burnout. *Journal of Occupational Health Psychology, 4*(4), 307–17.

Barnett, R. C., and Gareis, K. C. (2006). Role Theory Perspectives on Work and Family. In M. Pitt-Catsouphes, E. E. Kossek and S. Sweet (Eds.), *The Work and Family Handbook: Multi-disciplinary Perspectives and Approaches* (pp. 209–21). Mahwah, NJ: Lawrence Erlbaum.

Beutell, N. J., and Greenhaus, J. H. (1983). The Integration of Home and Nonhome Roles: Women's Conflict and Coping Behavior. *Journal of Applied Psychology, 68*(1), 43–48.

Bumpass, L. L. (1990). What's Happening to the Family? Interactions between Demographic and Institutional Change. *Demography, 27*(4), 483–98.

Burke, R. J. (1988). Some Antecedents and Consequences of Work–family Conflict. *Journal of Social Behavior and Personality, 3*(4), 287–302.

Cohen, A. (1997). Personal and Organizational Responses to Work/Non Work Interface as Related to Organizational Commitment. *Journal of Applied Social Psychology, 27*(12), 1085–114.

Crouter, A. C. (1984). Spillover from Family to Work: The Neglected Side of the Work–family Interface. *Human Relations, 37*(6), 425–42.

Duxbury, L. E., and Higgins, C. A. (1991). Gender Differences in Work–family Conflict. *Journal of Applied Psychology, 76*(1), 60–74.

Eby, L. T., Casper, W. J., Lockwood, A., Bordeaux, C., and Brinley, A. (2005). Work and Family Research in IO/OB: Content Analysis and Review of the Literature (1980–2002). *Journal of Vocational Behavior, 66*(1), 124–97.

Friedman, S. D., and Greenhaus, J. H. (2000). *Work and Family: Allies or Enemies? What Happens When Business Professionals Confront Life Choices.* New York: Oxford University Press.

Frone, M. R. (2003).Work–family Balance. In J. C. Quick and L. E. Tetrick (Eds.), *Handbook of Occupational Health Psychology* (pp. 143–62). Washington, DC: American Psychological Association.

Frone, M. R., Russell, M., and Cooper, M. L. (1992). Antecedents and Outcomes of Work Family Conflict: Testing a Model of the Work–family Interface. *Journal of Applied Psychology, 77*(1), 65–78.

Frone, M. R., Russell, M., and Cooper, M. L. (1997). Relation of Work–family Conflict to Health Outcomes: A Four-year Longitudinal Study of Employed Parents. *Journal of Occupational and Organizational Psychology, 70*(4), 325–35

Frone, M. R., Yardley, J. K., and Markel, K. S. (1997). Developing and Testing an Integrative Model of the Work–family Interface. *Journal of Vocational Behavior, 50*(2), 145–67.

Gattiker, U. E., and Larwood, L. (1990). Predictors for Career Achievement in the Corporate Hierarchy. *Human Relations, 43*(8), 703–26.

Greenhaus, J. H., and Beutell, N. J. (1985). Sources of Conflict between Work and Family Roles. *Academy of Management Review, 10*(1), 76–88.

Greenhaus, J. H., Collins, K. M., and Shaw, J. D. (2003). The Relation between Work–Family Balance and Quality of Life. *Journal of Vocational Behavior, 63*(3), 510–31.

Greenhaus, J. H., and Powell, G. N. (2006). When Work and Family Are Allies: A Theory of Work–family Enrichment. *Academy of Management Review, 31*(1), 72–92.

Greenhaus, J. H. (1988). The Intersection of Work and Family Roles: Individual, Interpersonal and Organizational Issues. *Journal of Social Behaviour and Personality. 3*(4), 23–44.

Grzywacz, J. G., and Butler, A. B. (2005). The Impact of Job Characteristics on Work–family Facilitation: Testing a Theory and Distinguishing a Construct. *Journal of Occupational Health Psychology, 10*(2), 97–109.

Gutek, B. A., Searle, S., and Klepa, L. (1991). Rational Versus Gender Role Explanations for Work–family Conflict. *Journal of Applied Psychology, 76*(4), 560–68.

Higgins, C. A., and Duxbury, L. E. (1992). Work–family Conflict: A Comparison of Dual Career and Traditional-career Men. *Journal of Organizational Behavior, 13*(4), 389–11.

Higgins, C. A., Duxbury, L. E., and Irving, R. H. (1992). Work–family Conflict in the Dual-earner Family. *Organizational Behavior and Human Decision Processes, 51*(1), 51–75.

Judge, T. A., Boudreau, J. W., and Bretz, R. D. (1994). Job and Life Attitudes of Male Executives. *Journal of Applied Psychology, 79*(5), 767–82.

Kinnunen, U., and Mauno, S. (1988). Antecedents and Outcomes of Work Family Conflict Among Employed Women and Men in Finland. *Human Relations 51*(2), 152–77.

Lerner, J. V. (1994). *Working Women and their Families*. Thousand Oaks, CA: Sage.

MacEwen, K. E., and Barling, J. (1994). Daily Consequences of Work Interference with Family and Family Interference with Work. *Work and Stress, 8*(3), 244–54.

Marks, S. R. (1977). Multiple Roles and Role Strain: Some Notes on Human Energy, Time and Commitment. *American Sociological Review, 39*(4), 567–78.

Marks, S. R., and MacDermid, S. M. (1996). Multiple Roles and the Self: A Theory of Role Balance. *Journal of Marriage and the Family, 58*(2), 417–32.

Mennino, S. F., and Brayfield, A. (2002). Job-family Trade-offs: The Multidimensional Effect of Gender. *Work and Occupations, 29*(2), 226–56.

O'Neil, R., and Greenberger, E. (1994). Patterns of Commitment to Work and Parenting: Implications for role Strain. *Journal of Marriage and the Family, 52*(1), 101–15.

Parasuraman, S., and Greenhaus, J. H. (2002). Toward Reducing Some Critical Gaps in Work–family Research. *Human Resource Management Review, 12*(3), 299–12.

Parasuraman, S., Purohit, Y. S., Godshalk, V. M., and Beutell, N. J. (1996). Work and Family Variables, Entrepreneurial Career Success, and Psychological Well-being. *Journal of Vocational Behavior, 48*(3), 275–300.

Parsons, T. and Bales, R. F. (1955). *Family, Socialization, and Interaction Process.* Glencoe, IL: The Free Press.

Pope, C. R. (1991). Effect of Domestic and Occupational Roles on Morbidity and Mortality. *Social Science and Medicine, 32*(7), 805–11.

Sieber, S. D. (1974). Toward a Theory of Role Accumulation. *American Sociological Review, 39*(4), 567–78.

Thomas, L. T., and Ganster, D. C. (1995). Impact of Family-supportive Work Variables on Work–family Conflict and Strain: A Control Perspective. *Journal of Applied Psychology, 80*(1), 6–15.

van Steenbergen, E. F., Ellemers, N., and Mooijaart, A. (2007). How Work and Family Can Facilitate Each Other: Distinct Types of Workfamily Facilitation and Outcomes for Women and Men. *Journal of Occupational Health Psychology, 12*(3), 279–300.

Voydanoff, P. (1988). Work Role Characteristics, Family Structure, and Work/Family Conflict. *Journal of Marriage and the Family, 50*(3), 749–61.

Waldron, I., Weiss, C. C., and Hughes, M.E. (1998). Interaction Effects of Multiple Roles on Women's Health. *Journal of Health and Social Behavior, 39*(3), 216–36.

Wayne, J. H., Randel, A. E., and Stevens, J. (2006). The Role of Identity and Work–family Support in Work–Family Enrichment and Its Work-related Consequences. *Journal of Vocational Behavior, 69*(3), 445–61.

Wiley, D. L. (1987). The Relationship Between Work/Nonworkrole Conflict and Job-related Outcomes: Some Unanticipated Findings. *Journal of Management, 13*(3), 467–72.

Williams, A., Franche, R. L., Ibrahim, S., Mustard, C. A., and Layton, F. R. (2006). Examining the Relationship between Work–Family Spillover and Sleep Quality. *Journal of Occupational Health Psychology, 11*(1), 27–37.

Weiss, R. S. (1990). Bringing Work Stress Home. In J. Eckenrode, and S. Gore (Eds.), *Stress between Work and Family* (pp. 17–38). New York: Plenum Press.

CHAPTER 10

Mindfulness Meditation: The Essential Stress-buster in Military Life Context

Nilanjana Sanyal

Professor and Former Head, Department of Psychology, University of Calcutta, West Bengal

Abstract The survival of the fittest is the demanding cry of the present world. The intricate dynamics of socio-political rivalry impacting the general flow of life, the self-centered materialism of human beings, the relentless competition in practical life, the ever-increasing pollution, the ever- growing population and resultant ecological imbalance, the health concerns due to virulent diseases have made this earth an enormously stressful domain to live in. Specific professions which involve frequent confrontations with hostile forces are crossing the limits of stress-bearing capacity. Hence, there is a dire need for an appropriate stress-buster. Mindfulness meditation seems to be a currently developed therapeutic mode to offer a rescue operation in this context. Overbearing stress is reducing logical fitness, curtailing emotional regulation, blurring our feelings to develop appropriate perspective in life. The resultant feature is restlessness, helpless state of self-inadequacy. To improve on our existential contour of realizational life, the primary purpose of mindfulness meditation is to cultivate a non-judgemental awareness of body and mind, and secondarily, to learn how to witness events and experiences on a moment-to-moment basis. Meditation practices can also help to foster insights into one's habitual and reactive patterns of perceiving and behaving, thus facilitating a desirable change in mental perspective. The elements of mindfulness meditation practice, like metacognition, mindful attention in the present moment experience, being non-judgemental, having re-perceiving capacity, having cognitive flexibility and developing meaning in life will be discussed in this chapter with their possible implications to combat the stress-load of military services. With its emphasis on cultivating tranquility, mindfulness meditation might seem like an odd fit for the military personnel.

The birth of a child into the world initiates the periphery of subjectivity. An immature child believes in self-pleasure only and promptly develops primary narcissism to draw pleasures in the poly morpho-perverse continuum. The external world is unknown to him owing to very limited cognitive strength and the volley of impulses coming from within the system which look for gratifications all the time. The hedonic contents in his mental sphere are only pleasure or displeasure. If and when his system is forced to encounter displeasure, frustration is the resultant feature. With days and years rolling by, when the child develops into a cognitive being, his understanding accommodates externality of the world with the concepts of objects therein. Amidst the context of interactions, his relationships evolve. Satisfactory flow of life contents make him feel contended, otherwise, the added-up frustrations result in stress.

The materialistic world of the 21st century with its advanced technological amenities and facilities has curved the desired patterns of its inmates in a way where intolerance of failure has made them look rigid in their perspective. People are mostly unaware of the fact that what they think, do and feel are the basic sources of their trouble because the flow of mental energy in the context is subjectivity-oriented demand patterns. Developing resilience in the advance face of tolerating stresses signifies maturity and non-suffering flairs in the life pattern. Psychology, as a behavioural science wing, tries to design and offer means and techniques which help humans to develop resilience to combat stress of various kinds. The ultimate aim there is to look for means of drawing satisfaction in life through graceful acceptance of reality and orienting one towards transcendence to have stress-free experience. Stress is a universal phenomenon. It is an evidence of life. A human being needs to be equipped enough psychologically to sail through life with ease and touch the fringes of existential bliss. Here is the offer of certain 'practice modes' or 'therapeutic tips' to be conscious of, to know and to practice, to develop or enhance quality resilience. One such means is 'Mindfulness Meditation,' an adaptive behavioural practice to reorient self from subjectivity to objectivity, to develop positive experiential perspective in understanding life in a global context of totality. The implications of the method will be verified in combating the usual stress-laden military services.

Stress in the Dynamics of Military Service

Of the multitude of services, military services belong to defence category. The role of military is to protect a whole nation and not an individual or a group. The enormity of this duty puts tremendous stress on military personnel. Obviously, the protector's role of not only a person, but a whole nation openly hints at the amount of stress being automatically associated with that. Being alert all the time, engaged in sharpening one's frustration tolerance limit, developing one's coping skill to combat stresses very often typically characterize the military pattern of services. Before joining the profession, personnel are aware of the stress factor; but once they are into the job the continuous exposure to stressful conditions pulls them down at times. Hence, it seems to be an important task of psychologists to chart stress-busters for these personnel, to free them off their negative emotion-loads, and be equipped to perform their responsible jobs.

Mental Health Advisory Team (MHAT) (2008) had clearly reported that in military service members, the usual demands of multiple deployments exposing the personnel to regular stressful environments have shown enormous deterioration in psychological and physical health conditions. The resultant effects included mental sufferings of the veterans, compromised effectiveness of the fighting force and high costs included in caring for combat veterans with stress-related ailments. While such stressors are widely recognized (Adler, McGurk, Stetz, and Bliese, 2003; King, King, Vogt, Knight, and Samper, 2006), comparatively less is known about effective methods for buffering against the adverse effects of such stresses as dysfunctional orientation and disease prevalence. In the context, a number of researchers (Grossman, Viemann, Schmidt, and Walach, 2004) have proved the efficacy of 'Mindfulness Training' (MT) as a stress-buster in a variety of civilian contexts. The same efficacy had been related to treating clinical populations as reported by Baer (2006). The MT protocols summarily attempts to set the mindfulness-mental mode being characterized by maximum attention to present-moment experiences, being devoid of judgement, elaboration or emotional reactivity (Stanley, Schaldach, Kiyonaga, and Jha, 2011). Mindfulness is a way of holding attention 'on purpose' (Chaskalson, 2011).

The Encapsulated Stress Context for Military Service

Even within the periphery of common-sense knowledge military service is an inherently stressful profession. Wearing the risk-taking responsibility of protecting a country from its predators on their sleeves, service members are openly exposed to traumatizing stressors before, during and after deployment, that include directly threats to individual safety, the need to inflict harm on others and encountering injury, death and significant human suffering (Adler *et al.*, 2003). Countless researchers suggest that troops are enveloped with considerable anxiety and distress in anticipation of deployment (Bolton, Litz, Britt, Adler and Roemer, 2001; MacDonald, Chaberlain, Long, Pereiran–Laird and Mirfin, 1998), resulting in possible higher risk for mental health problems after deployment. Recent researches unveil a strong association between mental health disorders and lower magnitude deployment stressors, including difficult living conditions, intense feeling of boredom and separation anguish from family (King *et al.,* 2006). Multiple deployments are found to earn a high toll, including lower morale, more mental health problems and more stress-related work problems (MHAT, 2008). Additionally, a range of negative health consequences, like post-traumatic stress disorder (PTSD) (Kaylor, King and King, 1987), depression (Erikson, Wolfe, King, King and Sharkansky, 2001), substance abuse (Boscarino, 1981) and physical health problems (Taft, Stern, King and King, 1999) are found to be the direct outcome of combat encounters.

The recent post-deployment conditions of the service members report a broad range of physiological disturbances like changes in sleep habits, low energy, headaches, chronic pain, cardio-pulmonary symptoms and gastro-intestinal difficulties (Levine, 2007; Scaer, 2008). The same condition was found to produce distinctive psychological disturbances including PTSD, traumatic brain injury, depression and anxiety disorders. In the same mental mosaic, PTSD was reported to have frequent comorbidity with other psychological problems and physical problems (Hoge, Kopian, Castro, Messer and Engel, 2007). The dysfunctions are frequently and generally labelled as independent issues that demanded situation-specific treatment. The process of analysis of such clinical conditions reveals that these disorders were the resultant effects of prolonged or extreme stress rather than the product of an illness with unrelated causes (Miliken, Auchterlonie and Hage, 2007; Tanielian and Jaycox, 2008; Hermann, 1992; Scaer, 2005; van der Kolk, Roth, Pelcovitz, Sunday and Spinazzoia, 2005). Since military service is crowning the glory of highest order protection for a country, this invaluable resource-repertoire needs to be served with all our power of behavioural reorientation. In the context, resilience means the ability to withstand strain and maintain healthy functioning even in highly stressful and demanding situations and to recover to full normalcy after the period of stress. A number of important studies suggest that resilience as a protective layer in individual personality can be cultivated or enhanced through mindfulness training (Feder, Nestler and Charney, 2009; Haglund, Nestadt, Cooper, Southwick and Charney, 2007). Hence, receiving training to cultivate resilience in the pre-deployment interval may directly counter the deleterious effects of the high-stress military context in tune of physical and psychological health. Thus, the beneficial effect of mindfulness training as a stress-buster is suggested for such profession.

The Beneficial Dynamism of Mindfulness Training

In eastern contemplative traditions, enhanced wellbeing is strongly associated with mindfulness training. In the West, interest in mindfulness has included efforts to define, measure and test mindfulness in order to develop a construct that connects mindfulness training with positive physical and mental health outcomes (Grossman *et al.*, 2004). Definitionally, the mindfulness training protocol points out the state of 'bringing one's attention to the present experience on a moment-by-moment basis.' (Marlatt and Kristeller, 1999, p. 68). Ample research evidences are available on the effectiveness of mindfulness-based interventions at reducing distress (Baer, 2003; Grossman *et al.*, 2004). The most well validated and known of such MT programme is mindfulness-based stress reduction (MBSR) (Kabat-Zinn, 1990). Recent studies of Carmody and Baer (2008) and Shapiro, Oman, Thoresen, Plante and Flinders (2008) subscribe to the same notion. Importantly, the common message from these studies is that a critical factor in determining the salutary effect of the procedure is the amount of spent time in mindfulness practice. The regular engagement with the process seems to be the most essential criterion. Recent reviews of resilience unveil the fact that important contributors to develop the power of resilience include willingness to face fears, the presence of positive emotions, the ability to attend to unpleasant stimuli, exposure to stressful experiences that are modestly destabilizing together with the capacity for cognitive flexibility, reappraisal, acceptance and emotion-regulation (Feder *et al.*, 2009; Haglund *et al.*, 2007). To unfold the true utilitarian value of mindfulness meditation in promoting resilience, the nature of the process, its constituting ingredients and their composite efficacy must be deduced for deeper understanding of the concept and establish its merits in handling stresses of various kinds.

History of the Concept of Mindfulness Meditation

According to Shapiro (1980), 'Meditation refers to a family of techniques which have in common a conscious attempt to focus attention in a non-analytical way and an attempt not to dwell on discursive, ruminating thought.' (p. 14). The important components of the process highlight the words, 'conscious' used explicitly to introduce the importance of the 'intention' to focus attention and the word 'attempt' which emphasizes 'process' rather than result. Moreover, the definition is found to be independent of religious framework or orientation. In fact, practices of mindfulness have their bases in positivistic wisdom traditions, but mindfulness is neither a religion nor a belief system. It is a skill like critical thinking that, once learned, becomes indispensable personality component.

Meditation techniques have been divided into concentrative meditation, mindfulness meditation (Goleman, 1972) and contemplative meditation (Kabat-Zinn, 1994). In all types of concentrative meditation, awareness is restricted by focusing attention on a single object. The attempt is to ignore other stimuli in the environment and focus complete attention on the object of meditation. The nature of focused attention here, is of a non-analytical, unemotional way, to directly experience the object of meditation either in the realm of external or the internal environment. It is an exercise in awareness which turns into the flow of breath. Mindfully focusing on the breath is like staring into a flame. We keep on, watching each breath from the bottom of the in-breath to the pause at the bottom of the out-breath. Examples of the object include the breath, as mantra, a single word like 'OM' (Benson and Proctor, 1984).

Mindfulness meditation is not about paying attention to the concept of the breath only. Rather, it is about following the physical breath with intense curiosity as to its texture, temperature and related sensations. These may include deep inhaling and exhaling through the nostrils as also the retention of the air in the lungs for longer intervals. The quality of attention we look for in mindfulness meditation is different from bearing down mentally. In such meditation, an attempt is made to attend non-judgmentally to all stimuli in the internal and external environment but not to get caught up in (ruminate on) any particular stimulus. Mindfulness meditation is referred to as an opening-up meditation practice. In fact, the first thing we observe when we practice mindfulness meditation is that the untrained mind is not at all cooperative. Instead of paying attention to each breath, the mind wanders in varied directions. It reaches into the future, worries about the past or judges meditation as a fruitless involvement. The failure to focus in the present moment can cause physical distractions too. But observing the wandering mind is the first step of mindfulness practice. And the way we observe the wandering mind is the key. A recent study has shown that such meditation is producing changes in the regions of the brain involved in learning and memory, regulation of one's emotions and self-awareness (McLean, 2012).

The third category is contemplative meditation. It involves opening and surrendering to a larger self (e.g., God, benevolent other), asking questions and bringing things unresolved. Such queries open up the vistas of new understandings and visions and prompt one to take new actions. Kabat-Zinn (1994) proposes that contemplative practices presuppose a certain degree of both concentration and mindfulness meditation skill that require opening up and focusing. To him, meditation is not a static category, but a dynamic process. Basically meditation is conceived as a developmental process, beginning with concentration and continuing with the ability to open up mentally and contemplate.

The primary purpose of all types of meditation is to cultivate a non-judgemental awareness of body and mind of the self, and secondarily to learn the technique of how to witness events and experiences on a moment-to-moment basis. Meditation practices can also help to improve one's habitual and reactive patterns of perceiving and behaving, thus initiating changes of these patterns where and when desired. It is to generate wisdom and compassion. Astin (1977) is of the opinion that mindfulness training provides powerful cognitive-behavioural coping tools. While having some commonness with other cognitive interventions, mindfulness-based approaches focus on attending to and altering cognitive processes rather than changing their content (Orsillo, Roemer, Block–Lerner, and Tull, 2004). Some authors have suggested that mindfulness training allows one to develop alternative paradigms and therefore interpret experiences in new ways, so that the crisis of a stressful situation may be perceived as an opportunity to progress further rather than a threat. Roemer and Orsillo (2002) call this 'cognitive flexibility'. Mindfulness-based stress reduction (MBSR) is also thought to cultivate self-regulation, being related to positive changes in both physical and psychological health (Coffey and Hartman, 2008). This 'self-regulated' psychological orientation is used to describe a kind of learning that is guided by metacognition, strategic action and motivation to learn. People who have developed their self-regulatory capacities as a product of MBSR are better able to calibrate their emotions, and they tend to attribute their successes or failures to factors within their control. They usually exhibit a high sense of self-efficacy (Pintrich and Schunk, 2002). Thus research-based evidences prove that meditation techniques, including MBSR, can cultivate qualities such as compassion, forgiveness, mindfulness and spirituality inviting the positive psychological components in one's behavioural frame. In the context, Kabat-Zinn (1990, 2003) outlined seven foundations of mindfulness practice within the Positive Psychology fold.

1. *Non-judging:* Being aware of judgemental qualities and perceiving inner and outer experiences;
2. *Patience:* Understanding and accepting that sometimes things will unfold in their own time;
3. *Beginner's mind:* Observing everything as if for the first time;
4. *Trust:* Taking responsibilities for being oneself and learning to listen to and trust one's own being;
5. *Non-striving:* Realizing that there is no goal other than searching for self;

6. *Acceptance:* experiencing things as they actually are in the present; and
7. *Letting Go:* Releasing thoughts, feelings and situations that the mind seems to be enveloped with.

The Process of Imbibing Mindfulness Qualities

All meditation disciplines centre on the practice of cultivating attention. However, the 'intention' with which 'attention' is directed, may be significant in maximizing optimal health. By making intention explicit in meditation research and practice, it is well assumed that positive psychological qualities may be enhanced. Intension is conceived to be composed of two elements:

1. The quality of attention designed to as 'mindfulness qualities', and;
2. The content of attention framed as 'systemic perspectives'.

The Intentional Systemic Mindfulness takes into account both the 'nature of the attention' through the mindfulness qualities and the 'framework' within which attention is initiated through the systemic perspectives. In short, it verifies two critical aspects of intention—how we attend and why we attend bringing out the nature of mindfulness practice together with its implications in our conscious life. All meditation techniques are grounded on the cultivation of attention. Mere 'paying attention' is not enough, how one attends is important as it can be responsible for cultivating health in myriad dimensions. On the other hand, if someone focuses his attention with a conscious intention to blend attention with mindfulness qualities of acceptance, generosity and fairness, this intention may indeed be the hallmark of health-promoting tendency.

Twelve mindfulness qualities are intentionally incorporated and present in conscious attention. Seven were initially defined by Kabat-Zinn (1990) as mentioned earlier with an addition of five being mentioned by Shapiro and Schwartz (2000) which are as follows:

1. *Gentleness:* Characterized by a soft, considerate and tender quality; not really being passive, indisciplined or indulgent in character orientation.
2. *Generosity:* Gaining in the present moment within a context of love and compassion, but having detachment orientation in the context.
3. *Empathy:* The quality of feeling and understanding another person's present moment life context including his/ her emotions and reactions and communicating this to the person.
4. *Gratitude:* The quality of reverence, appreciation and thankful acknowledgement for the present moment.
5. *Loving kindness:* A quality signifying benevolence, compassion and cherishing a quality of having deep marks of forgiveness and unconditional love.

The set of qualities of mindfulness meditation implies that by incorporating these qualities within one's mental system, people can really move out of the just reverse negative qualities, the accumulation of which changes our perspectives in life and bestows on us loads of stressful reactions. Hence, as an answer to stress-buster, one of the techniques of the process is the development of greater understanding through the systemic cultivation of inquiry and insight.

The Ingredients of Mindfulness Meditation: The Invited Positive Elements of Mind

The mindfulness meditation practice necessarily hints at the developmental folds of maturity-attainment process. The self-system being exposed to the materialistic competitive world is bound to incorporate negative components of mind like jealousy, non-cooperation, competitiveness, sense of inadequacy, senses of deficiency or deprivations in life. All these together compose the negative life perspective calling forth rigidity in the self-system, having self-centeredness and non-empathic personality mould. The life flow in socio-emotional parlance keeps staggering, affecting the basic senses of cohesion and cooperation. We all suffer in such an ambience and our good senses are one of the inherent features which at times prompt us to find out means of avoiding these negativities and invite positive components in the same context. Mindfulness meditation seems to be the mental key to open up self to itself and become aware of such follies within us. The very fact of becoming aware of our killing impulsive processes that are wrapped up in immaturity, enveloping the true original self of faith, trust and mutuality, will automatically drop off the negative cast and will be replaced by better set of positive features of the self. Delving into the process-work of mindfulness meditation is expected to bring out in the light its constitutional ingredients as follows in the free-flowing wandering mind of thoughts and feelings.

1. Metacognition

The capacity to know what we are thinking, feeling, or sensing as it is going on is what might be called the 'metacognitive' dimension of mindfulness. Metacognition refers to our insightful knowledge about our own cognitive processes or anything related to them (Flavell, 1976). In the context of mindfulness, metacognition extends into the domains of feelings and body-sensations of the present moment. All of these, thoughts, feelings and body sensations are experienced in thinking, 'that' it is feeling, 'that' it is sensing. Anyone of these metacognitive elements of mindfulness can take place in our experience at any time and transform it. We are able to make wiser choices then. In fact, one of the purposes of mindfulness practice is to significantly increase the chance of that happening.

2. Existential Evidence

Another quality of mindfulness attentive process, according to Kabat-Zinn's (1990) definition, is that it is rooted 'in the present-moment', that is phenomenologically existential evidence. Generally our attention, other than being directed at something present or in fore of us, is oriented towards the future or the past. Our future seems unknown and hence thoughts regarding future occurrences will either be fantasized positively or else anxiety ordered regarding some unwanted components. Things are not just within our control there. Similarly, if the mind keeps on tracing back the past evidences, either of positive or negative quality, we seem to have no control over the felt experiences. The harping of mind tries to revisit the older situations in life, feeling either happy or morose about them. But present experiences together with their clear consciousness are energizing in one sense. Counting on our failures or deficiencies may produce rigid morality that tightens up our system and we fail to touch upon the present 'bliss'. Conscious present experiences with the touch of the magic wand of positive perspective within us may provide us the opportunities to refreshen ourselves and broaden our emotional horizons. Rather than having the wandering mind which is always preoccupied by the past or future, the fresh consciousness of the immediate present seems to rejuvenate us with fresh positive angles of experience that is desired in our present study-quality. Moyers (1993), citing Jha, Krompinger and Baime (2007), thus grins, 'to place your attention where you want it to be and to keep it there for longer is a known outcome of mindfulness training' (Moyers, 1993, p. 18).

Jha and Stanley (2010) had shown in their study that just as daily physical exercise leads to physical fitness, being involved in mindfulness exercises on a regular basis improved the Marines' mind-fitness by extending their working memory under stress. That, they claimed, safeguards then against distraction and emotional reactivity and lets them maintain a mental work-parlance that ensures quick decisions and action plans. Besides offering some protection to combatants from post-traumatic stress and other anxiety disorders, the mindfulness training opened the vistas of clear thinking needed for soldiers fighting in challenging and ambiguous counter-insurgency zones. For Jha and Stanley, the study showed that mindfulness training help in maintaining peak performance in the face of extremely stressful circumstances: the emergency services, relief workers, trauma surgeons, professional and Olympic athletes, and so on. Mindfulness meditation seems to produce an increased capacity to deploy their attention in ways that their tasks require.

3. Non-judgemental Mental Fold

The quality of mindful awareness that Kabat-Zinn (1990) draws attention to is that it is 'non-judgemental'. This does not mean that one does not make judgements or that one does give up the powers of discrimination. Far from it, it says that it indicates our frequent tendency of constantly judging ourselves in a critical light. The judgemental orientation there signifies once again the rigidity in the self, curbing the power of realistic acceptance and widened accommodativeness. It seems to be a kind of irrational tyranny that can never be satisfied (Williams, Teasdale, Segal and Kabat-Zinn, 2007). The mindfulness approach, by contrast, is to let one experience what one is experiencing without censoring it, without blocking things in order to have only right kind of things in the mind. Mindfulness training prompts people to bring an attitude of warm, kind curiosity to their direct experiences—encompassing thoughts, feelings and body-sensations from moment to moment. Learning to pay attention in this way—on purpose, in the present moment and non-judgementally—participants in mindfulness courses begin to experience a radical shift in perspective. Hence, clarity and objectivity of personal present experiences become the outcome of mindfulness training.

4. Re-perceiving

The last ingredient of mindfulness meditation seems to be the product of shifted perspective due to its process, that is 're-perceiving' (Shapiro, Carlson, Astin and Freedman, 2006). Rather than being emotionally entangled in the drama of their personal narratives or life stories, participants on MBSR courses learn the skill of taking a back foot and witnessing what is going on for them. This skill of re-perceiving brings them the skill of converting 'subjectivity' into 'objectivity' (Goleman, 1980). The shift is in fact wisdom-orienting detachment process of self from subjectively enmeshed emotions of daily chores. Such a shift is conceived as an 'approach' mode of mind, characterized by warmth, acceptance and curiosity, the elements of Positive Psychology that prompt effective handling of any situation on the part of the experiencing partner. 'Re-perceiving' is not the same as detachment. It is not about distancing oneself from one's experience to the point of apathy or numbness. Instead, the experience of mindful re-perceiving gives rise to a deep knowledge regarding present experience. Mindfulness allows for a degree of distance from one's experience in the sense that one becomes clearer about what one is experiencing. But, this does not result in disconnection or disassociation. Instead, it allows one to experience the changing flow of one's mental and physical experiences without getting identified with them and getting fixated to them. Ultimately, the outcome seems to bear a profound, penetrative, non-conceptual knowledge into the nature of mind and world (Kabat-Zinn, 2003).

This capacity to re-perceive, although it needs to be learned, is something people can consciously train in and it is simply a continuation of the way we naturally develop. In our developmental trajectory of life, we develop increasing capacity for objectivity about our internal experiences. In psychodynamic language this is our 'object relation' and attainment of 'object constancy' as unconscious processes. Such developments initiate the development of the ability to 're-perceive'. The re-perceptive process generates 'empathy' within us and ultimately widens the periphery of self-growth. Mindfulness practice continues this natural process, but now at the conscious level. With mindfulness training, we can consciously develop an increasing capacity for objectivity with regard to our internal and external experiences (Shapiro *et al.*, 2006) and become personally responsible persons. Etches of maturation set their marks on our life canvas and mental reflections. The conversion from unmindfulness to mindfulness—from unconsciously clinging to the state of intimate objectivity around that experience—is almost alchemical in its subtlety. A kind of transmutation takes place that allows the recognition and understanding that what was once threatening or compulsively desirable can become much more tolerable and effectively manageable.

The Need for Improving Our Existential Contours of Realizational Life

Life undoubtedly has concrete, realistic bases. The sensory-experiential contents fulfil or frustrate our hedonic impulses. We have sense of happiness and are given to fits of anger, fear, and worry almost alternately. No one can have seamless peaceful existence. The existential evidences are marked with both positivities as well as negativities. Maturity orientation puts its radar on widened acceptance spectrum to accept reality as gracefully as possible. The task is not that simple, but demands awareness on our part to realize the fact that we are thrown into the abyss of existence to experience everything that we set into the realm of authenticity, our realizational folds change in the direction of making us equipped to deal with our stresses of various kinds. Understanding and feeling components prompt us to alter our realizational contours, but we need to adopt appropriate actions to accomplish this. Mindfulness meditation, in the theoretical folds of analysis together with research evidences, has framed the way for us to follow a system of developing enough resilience to combat negative experiences of myriad magnitudes. Following the system, there seems to be bright expectations that our realizational fold of experience will emit the fragrance of a desirable quality to tread on context of living with greater vigour and required humility.

Conclusion

The content called 'life', the context called 'living', inevitably will have tumultuous evidences of positivities and negatives. Stresses are challenges in life, that can propel the attempt to accept them in positive mind-frame or else, they are craters of darkness that initiate helplessness, hopelessness and depression in negative mental flickers. Stressful service contexts like military profession demand enough strength of mind together with physical fitness to carry the unwanted loads of each and every deployment. For this they need to be trained. One such training is mindfulness meditation. Defense personnel are assets of our socio-political life. The badges on their pockets and sleeves dazzle with achievement glories. But, after all they are human beings, with their share of personal frailties. Why not help them with techniques, so that they can overcome their breakdowns?

References

Adler, A. B., McGurk, D., Stetz, M. C., and Bliese, P. D. (2003). *Military Occupational Stresses in Garrison, Training, and Deployed Environments.* Paper Presented at the National Institute for Occupational Safety and Health, American Psychological Association Symposium, Toronto, Canada.

Astin, J. A. (1997). Stress Reduction through Mindfulness Meditation: Effects on Psychological Symptomatology, Sense of Control and Spiritual Experiences. *Psychotherapy and Psychosomatics, 66*(2), 97–106.

Baer, R. A. (2003). Mindfulness Training as a Clinical Intervention: A Conceptual and Empirical Review. *Clinical Psychology Science and Practice, 10*(2), 125–43.

Baer, R. A. (2006). *Mindfulness-based Treatment Approaches: Clinician's Guide to Evidence Base and Applications.* Boston: Elsevier, Academic Press.

Benson, H., and Proctor, W. (1984). *Beyond the Relaxation Response.* New York: Putnam/Berkeley.

Bolton, E., Litz, B., Britt, T., Adler, A., and Roemer, L. (2001). Reports of Prior Exposure to Potentially Traumatic Events and PTSD in Troops Poised for Deployment. *Journal of Traumatic Stress, 14*(1), 249–56.

Boscarino, J. (1981). Current Excessive Drinking among Vietnam Veterans: A Comparison with Other Veterans and Non-veterans. *International Journal of Social Psychiatry, 27*(3), 204–12.

Carmody, J. and Baer, R. A. (2008). Relationships Between Mindfulness Practice and Levels of Mindfulness, Medical and Psychological Symptoms and Well-being in a Mindfulness-based Stress Reduction Programme. *Journal of Behavioural Medicine, 31*(1), 23–33.

Chaskalson, M. (2011). *The Mindful Workplace: Developing Resilient Individuals and Resonant Organizations with MBSR.* Chichester, West Susex: John Wiley & Sons, Ltd.

Coffey, K. A., and Hartman, M. (2008). Mechanisms of Action in the Inverse Relationship Between Mindfulness and Psychological Distress. *Complementary Health Practice Review, 13*(2), 79–91.

Erikson, D., Wolfe, J., King, D., King, L., and Sharkansky, E. (2001). Post-traumatic Stress Disorder and Depression Symptomatology in a Sample of Gulf War Veterans: A Prospective Analysis. *Journal of Consulting and Clinical Psychology, 69*(1), 41–49.

Feder, A., Nestler, E. J., and Charney, D. S. (2009). Psycho-biology and Molecular Genetics of Resilience. *Nature Reviews Neuroscience, 10*(6), 446–57.

Flavell, J. H. (1976). Meta-cognitive Aspects of Problem-solving. In L. B. Resnick (Ed.), *The Nature of Intelligence* (pp. 232–36). Hillsdale, NJ: Erlbaum.

Goleman, D. (1972). The Buddha on Meditation and States of Consciousness: Part II. A Typology of Meditation Technique. *Journal of Transpersonal Psychology, 4*(2), 151–210.

Goleman, D. (1980). A Map for Inner Space. In R. N. Walsh and F. Vaughan (Eds.), *Beyond Ego* (pp. 141–150). Los Angeles: J. P. Tarcher.

Grossman, P., Viemann, L., Schmidt, S., and Walach, H. (2004). Mindfulness-based Stress Reduction and Health Benefits: A Meta-analysis. *Journal of Psychosomatic Research, 57*(1), 35–43.

Haglund, M. E. M., Nestadt, P. S., Cooper, N. S., Southwick, S. M., and Charney, D. S. (2007). Psychobiological Mechanisms of Resilience: Relevance to Prevention and Treatment of Stress-related Psychopathology. *Development and Psychopathology, 19*(3), 889–90.

Hermann, J. L. (1992). Complex PTSD: A Syndrome in Survivors of Prolonged and Repeated Trauma. *Journal of Traumatic Stress, 5*(3), 377–91.

Hoge, C. W., Kopian, A. T., Castro, C. A., Messer, S. C., and Engel, C. C. (2007). Association of Post-traumatic Stress Disorder with Somatic Symptoms, Health Care Visits and Absenteeism among Iraq War Veterans. *American Journal of Psychiatry, 164*(1), 150–53.

Jha, A. P., Krompinger, J., Baime, M. J. (2007). Mindfulness Training Modifies Subsystems of Attention. *Cognitive, Affective and Behavioral Neuroscience, 7*(2), 109–19.

Jha, A. P., and Stanley, E. A. (2010). Examining the Protective Effects of Mindfulness Training on Working Memory Capacity and Affective Experience. *Emotion, 10*(1), 54–64.

Kabat-Zinn, J. (1990). *Full Catastrophe Living.* New York: Delta Books.

Kabat-Zinn, J. (1994). *Wherever You Go, There You Are.* New York: Hyperion.

Kabat-Zinn, J. (2003). Mindfulness-based Interventions in Context: Past, Present and Future. *Clinical Psychology: Science and Practice, 10*(2), 144–56.

Kaylor, J., King, D., and King., L. (1987). Psychological Effects of Military Service in Vietnam: A Meta-analysis. *Psychological Bulletin, 102*(2), 257–71.

King, L., King, D., Vogt, D., Knight, J., and Samper, R. (2006). Deployment Risk and Resilience Inventory: A Collection of Measures for Studying Deployment-related Experiences of Military Personnel and Veterans. *Military Psychology, 18*(2), 89–120.

Levin, A. (2007). Multiple Physical Illnesses Common in Iraq War Veterans with PTSD. *Psychiatric News, 42*(2), 4–5.

MacDonald, C., Chaberlain, K., Long, N., Pereiran-Laird, J., and Mirfin, K. (1998). Mental Health, Physical Health, and Stressors Reported by New Zealand Defense Force Peacekeepers: A Longitudinal Study. *Military Medicine, 163*(7), 477–81.

Marlatt, G. A., and Kristellar, J. L. (1999). Mindfulness and Meditation. In W. R. Miller (Ed.), *Integrating Spirituality into Treatment* (pp. 67–84). Washington, DC: American Psychological Association.

Moyers, B. (1993). *Healing and the Mind.* New York: Broadway Books.

McLean, V. A. (2012, May 08). A New Component of US Military Combat Preparedness: Meditation. *Veterans Today.*

Mental Health Advisory Team (MHAT) (2008). *Operation Iraqi Freedom, 06–08: Iraq Operations Enduring Freedom 8: Afghanistan.* The United States of America: Department of Defense.

Miliken, C. S., Auchterlonie, J. L., and Hage, C. W. (2007). Longitudinal Assessment of Mental Health Problems Among Active and Reserve Component Soldiers Returning from the Iraq War. *Journal of the American Medical Association, 298*(11), 2141–48.

Orsillo, S. M., Roemer, L., Block-Lerner, J., and Tull, M. T. (2004). Acceptance, Mindfulness and Cognitive-Behaviour Therapy: Comparisons, Contrasts and Applications to Anxiety. In S. Hayes, M. Linchan and V. Folette (Eds.), *Mindfulness and Acceptance: Expanding the Cognitive-behavioural Tradition* (pp. 66–95). New York: Guilford.

Pintrich, P. R. and Schunk, D. (2002). *Motivation in Education: Theory, Research and Applications* (2nd Ed.). Upper Saddle River, NJ: Merrill Prentice Hall.

Roemer, L. and Orsillo, S. M. (2002). Expanding our Conceptualization of and Treatment for Generalized Anxiety Disorder: Integrating Mindfulness/Acceptance-based Approaches with Existing Cognitive-Behavioural Models. *Clinical Psychology: Science and Practice, 9*(1), 54–68.

Scaer, R. (2005). *The Trauma Spectrum.* New York: W. W. Norton.

Scaer, R. (2008). *Trauma, Dissociation and the Healing of Combat Stress.* Presentation to the Conference of the National Institute for the Clinical Application of Behavioural Medicine. Hilton Head Islands, South Carolina. Mp3 Recording Retrieved from: http://www.iplayback.com/prod_detail_list/166/2

Shapiro, D. H. (1980). *Meditation: Self-regulation Strategy and Altered State of Consciousness.* New York: Aldine.

Shapiro, S. L., and Schwartz, G. E. (2000). The Role of Intention in Self-regulation: Toward Intentional Systemic Mindfulness. In M. Boekaerts, P. R. Pintrich and M. Zeidner (Eds.), *Handbook of Self-regulation* (pp. 253–73). New York: Academic Press.

Shapiro, S. L., Carlson, L. E., Astin, J. A., and Freedman, B. (2006). Mechanisms of Mindfulness. *Journal of Clinical Psychology, 62*(3), 373–86.

Shapiro, S., Oman, D., Thoresen, C., Plante, T., and Flinders, T. (2008). Cultivating Mindfulness: Effects on Well-being. *Journal of Clinical Psychology, 64*(7), 840–62.

Stanley, E. A., Schaldach, J. M., Kiyonaga, A., and Jha, A. P. (2011). Mindfulness-based Mind Fitness Training: A Case Study of a High-stress Predeployment Military Cohort. *Cognitive and Behavioral Practice, 18*(4), 566–76.

Taft, C., Stern, A., King, L., and King, D. (1999). Modeling Physical Health and Functional Health Status: The Role of Combat Exposure. Post-traumatic Stress Disorder and Personal Resource Attributes. *Journal of Traumatic Stress, 12*(1), 3–23.

Tanielian, T. and Jaycox, L. H. (2008). *Invisible Wounds of War: Psychological and Cognitive Injuries, their Consequences and Services to Assist Recovery.* Santa Monica, CA: Centre for Military Health Policy Research, RAND Corporation.

van der Kolk, B. A., Roth, S., Pelcovitz, D., Sunday, S., and Spinazzoia, J. (2005). Disorders of Extreme Stress: The Empirical Foundation of a Complex Adaptation to Trauma. *Journal of Traumatic Stress, 18*(5), 389–99.

Williams, M., Teasdale, J., Segal, S., and Kabat-Zinn, J. (2007). *The Mindful Way Through Depression: Freeing Yourself from Chronic Unhappiness.* London: Guildford Press.

CHAPTER 11

Resilience: Relevance to Military Context

Rajbir Singh and Lokesh Gupta

Maharishi Dayanand University, Rohtak, Haryana

Abstract Life is full of positives and negatives, some of them being moderate while others a few are intense. When someone fails to reconcile and manage severe negative experiences, it takes him to the blues. However, many of us sail through and again come to the normal track of life to face further naturally occurring challenges. Previous severe negative experiences help us to bounce back for positive adaptation in the context of significant adversity or risk. They are now better psychologically prepared and can call upon particular resources when facing risk and disadvantages. Present-day work context is highly challenging and full of social conflict and not even free from trauma caused by unexpected hazards. People who display effective coping mastery and positive adaptation tend to be resilient. The concept of psychological resilience has been linked to not only recovery and sustained purpose, but that of growth too. The construct is expanding from personal psychological resilience to collective psychological resilience. It is, therefore, expected that resilient group/work force shall be capable to coordinate and draw upon collective resources to deal with adversity. Positive Military psychology has gained acceptance amidst great excitement. It has helped in moulding soldiers to be successful and resilient. Military operation, long term or short term, are full of risks and challenges which naturally creates a need to think of outcomes and prepare for its bipolar variants. In military environment, being highly strategic and counter strategic, oddities are not uncommon. Therefore, there should be specific programmes to build resilience at the personal and group levels by evolving models from clinically successful cases where the resilient patients have been found to follow proper treatment regimen, better prognostics, and future control and even extended life expectancies. In work context, several benefits from industrial and organizational psychology have come up when hardiness training has been effective. The concept of salutogenic hardiness and combination of certain positive effects associated with character strengths are considered imperative by certain resilience building models. Although it shall be a paradigm shift and shall also require a high correlated character strengths including personality are factors combined to develop resilience building models. Although it shall be a paradigm shift and shall also require a high correlated character strengths including personality are factors combined to develop resilience building models. Although it shall be a paradigm shift and shall also require a high level decision in military organizations, some initial affirmative experiences and boldness to change a mindset would facilitate the adoption of resilience-building models.

While asking the lay meaning of the term 'Resilience' from a large group of Indian Army officers during a training programme, it was observed that majority of them (80%) understand it in terms of bouncing back and effective coping (68%). However, when asked to choose among the several conventional meanings (for example, 'Is conditional to adversity'; 'Ability to bounce back'; 'Capacity for adapting psychologically, emotionally, physically', 'Post traumatic growth', 'Effective coping', and 'Is a trait, an ability and a skill', taken as alternatives for the meaning of resilience), 'ability to bounce back' was the most popular, followed by 'effective coping' at the second place and 'capacity for adapting emotionally' at the third place.

Life is full of positives and negatives, some of them being moderate while a few severe. Making adjustments in each life stage, coping with unexpected setbacks, or handling the daily stresses of life can turn a crisis into an opportunity for growth. Resilience, as described by American Psychological Association (2011), is similar to taking a raft trip down a river.

> On a river, you may encounter rapids, turns, slow water and shallows. As in life, the changes you experience, affect you differently along the way. In travelling the river, it helps to have knowledge about it and past experience in dealing with it. Your journey should be guided by a plan, a strategy that you consider likely to work well for you. Perseverance and trust in your ability to work your way around boulders and other obstacles are important. You can gain courage and insight by successfully navigating your way through the water. Trusted companions who accompany you on the journey can be especially helpful for dealing with rapids, upstream currents, and other difficult stretches of the river. You can climb out to rest alongside the river. But, to get to the end of your journey, you need to get back in the raft and continue
>
> (American Psychological Association, 2011, p. 8)

Psychologists have been studying resilience since the 1970s and research has demonstrated that there are many aspects to resilience (Reivich and Shatte, 2002; Seligman, 1990). Masten (2001) defines resilience as a set of processes that enable good outcomes in spite of serious threats.

In recent years, the discourses on 'resilience' have emerged in which elements formerly identified as human 'attributes' such as courage, will power, fortitude and character have been reconfigured as 'coping strategies' or 'skills' that can be learned by anyone.

> Resilience implies a systematic, widespread, organizational, structural and personal strengthening of subjective and material arrangements so as to be better anticipated and tolerate disturbances in a complex world without collapse, to withstand shocks, and to rebuild as necessary... a logic of resiliency would aspire to create a subjective and systematic state to enable each and all to live freely and with confidence in a world of potential risks
>
> (Lentzos and Rose, 2009, p. 243).

Recovery, sustained purpose and growth are three cardinal features of resilience. When stress disturbs the pace of ongoing performance, many quickly restore the previous level and even sometimes better than the earlier, that marks the *recovery* of function. More so, in this process, the first priority becomes the safety, survival and conservation of resources which altogether shifts the focus from the purpose; however, resilience helps to *sustain the purpose* despite temporary shift. Passing through these initial troubles furthers the *growth* of the person enabling him to tackle the same stressor with ease next time.

Resilience is considered as a process, which involves many systems from cells to individuals to families to societies. An individual may be said to have more or less capacity for resilience, but the actual pattern of an individual's behaviour results from many interactions, both within the person and then between the person and environment. Because of the involvement of many interactions resilience cannot be viewed as a single trait.

Many definitions on resilience consider that it is an ability to bounce back from adversity, positively adapting to adverse conditions and function well even while facing unexpected harmful/adverse situations. Resilience covers many concepts like awareness of personal resources, effective communication with family and friends, detection of threat, reaction to adversity (sometime resistance) and recovery of pre-event functioning. Resilience is a hypothetical concept coupled with situation. We can judge resilience in terms of adaptive behaviour of an individual with positive outcomes of risk/adversity. After all, resilience is a never-ending process in life and adversity is the hard reality of life. The term 'resilience' had been in use for long in material sciences, environmental sciences and even economics (resilient buildings, resilient ecology, resilient economy, etc.). Quite frequently the term resilience is prefixed with 'psychological' and thereby emphasizing that it is a psychological phenomenon—while referring to human beings. However, its description has recently been extended from individual to social/collective domain of behaviour. More often, we have been talking of resilient communities. Psychological resilience has two levels: Personal/Individual and Collective Psychological Resilience [Organizational and Community in McAslan's (2010) description].

Personal Psychological Resilience is the key for creating a dynamic culture and having people who are 'fit' and ready to seize new opportunities and respond to the demands of business both now and in future. It is a sort of psychological preparedness. Personal resilience is a multifaceted concept, and it includes mental level of resilience (self-confidence, desire to achieve, mental toughness, etc.), emotional ability to be resilient (emotional intelligence), physiological ability of resilience (proper diet, exercise, relaxation, quality sleep, etc.) and individual personality (purpose/meaning of life, hardiness, etc.). Norris, Tracy and Galea (2009) suggest that resilience represents just one of a number of categories of reaction by adults (that is, resistance, resilience, recovery, relapsing/remitting, delayed dysfunction and chronic dysfunction) following exposure to trauma or severe stress.

Collective Psychological Resilience refers to the way in which groups or crowds of people 'express and expect solidarity and cohesion, and thereby coordinate and draw upon collective sources of support and other practical resources adaptively to deal with adversity' (Drury, 2009, p. 21). Collective psychological resilience covers all aspects of personal psychological resilience and recovery in term of mass emergencies (for example, flood, earthquake, war, thunder, tsunami, terrorist attack, etc.). It also includes teamwork, leadership, social norms, group cohesiveness, etc. The people around us make a protective shield and reduce the risk of harm. Understanding the psychological resilience at the level of community is not simple as the process implies the interplay of individuals, families, group as well as the physical environment. Different theorists have offered to enhance resilience of communities (for example, Adger, 2003; Paton and Johnston, 2006; Pendall, Foster and Cowell, 2007). Some protective factors have been suggested as frequent and supportive interactions, shared norms and values, better information and adaptability at the community level (Mayunga, 2007; Norris, Stevens, Pfefferbaum, Wyche and Pfefferbaum, 2007). The military is working like a community and work for the welfare of communities.

Resilience Framework in Military Context

This chapter emphasizes on military context which essentially is the interplay of individual (soldier), group (unit or battalion) and communities (families). It therefore, falls under collective psychological resilience and extends beyond military context during wartime. Organization of various types altogether forms communities which imply resilient organizations, that is, resilient community. Resilience in organizations is ensured by incorporating flexibility in its members, adaptability to shifting needs of materials and strong structure. Seville *et al.* (2006) suggest that a resilient organization is one that is 'able to achieve its core objectives in the face of adversity. This means not only reducing the size and frequency of crisis (vulnerability), but also improving the ability and speed of the organization to manage crisis effectively (adaptive capacity). To effectively manage the crisis, organizations also need to recognize and evolve in response to the complex system within which the organization operates (situational awareness) and to seek out new opportunities even in times of crisis' (Seville *et al.*, 2006, p. 4). Two such attempts shall be illustrative of resilience pursuits in military context. In the United States, a comprehensive management framework has been provided for organizational resilience. It comprises anticipation, prevention, preparation and responding to disruptive event. However, in the United Kingdom, a broader framework includes analysis of critical national infrastructure, chemical and biological threats, transport security, intelligence and counter-terrorism, border force and immigration for national security and resilience. Thus, the construct of psychological resilience can be extended from individual through communities (organization) to nations, even globally.

Kumpfer (1999) has developed the resilience framework comprising process and outcome construct. Modelling under such a framework, he applied it extensively on youths in the context of drug addiction and de-addiction with positive outcomes. There are six major constructs in the model (four influence domains and two transactional points). There seems to be an appeal in the model for its application in military context. Some of the processes of resilience is not bounded by the static factors of the organization or context, the acute stressor or challenge, the environ-

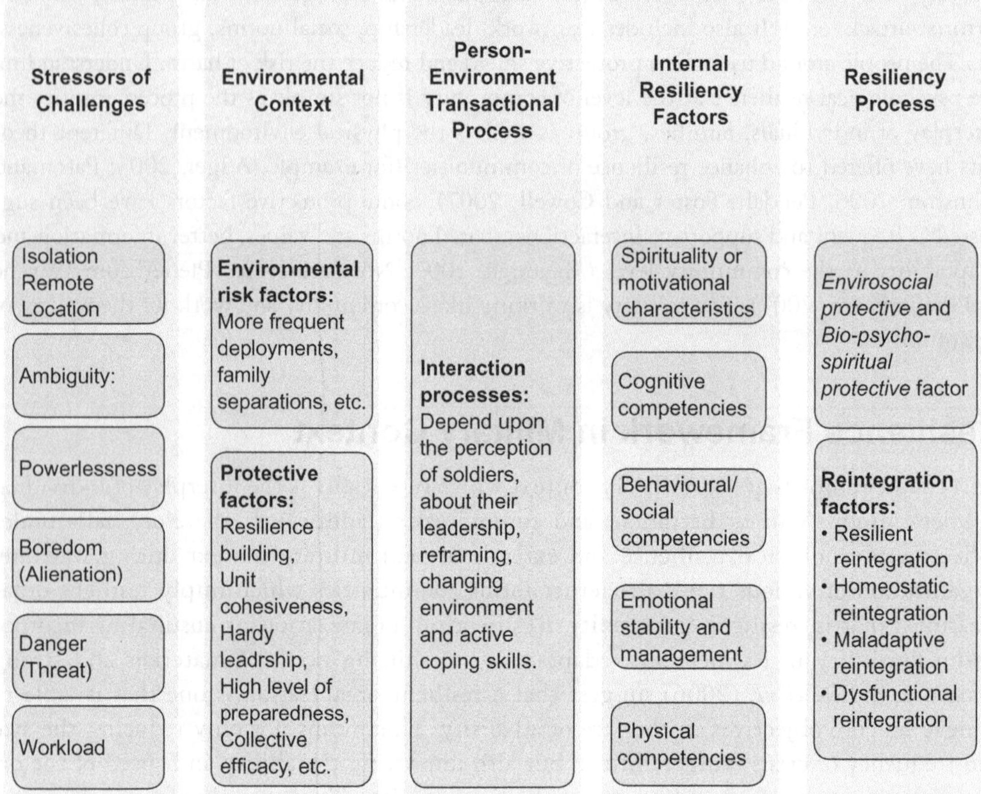

Figure 11.1 Adaptive Version of the Kumpfer (1999) Resilience Model.

mental context, the individual characteristics and the outcome. For this purpose, the model has been simplified and suitably adapted (Figure 11.1). The description of influence domains supplemented with military context follows next.

1. Stressors or Challenges

The resilience framework or processes begin with an initial event and ends with an outcome. Stimulus in resilience situation should be some type of stressors or challenges. Challenges help a person to face new stressors and to grow from the experience. This is the essence of resilience. In the military setting, there are so many stressors and challenges like high ambiguity, unpredictable enemy, diffused threat, powerlessness (Delahaij, Gaillard and Soeters, 2006; Stetz *et al.*, 2007; Sharma and Sharma, 2008). There has been addition of newer type of stressor/challenges in recent wars, for example, Iraq war, Afghanistan war, global terrorism, low-intensity prolonged conflict at the borders between several nations, coalition defence forces, peace-keeping forces, and so forth. Soldiers are deployed in newer far-off territories with cultural and linguistic barriers. Work units are configured with unknown colleagues from different nationalities.

Quite often, the command is fractured and the purpose is not disclosed. Soldiers need to overcome these confusions, work hard to build trust and then survive in hazardous warzone. Prolonged cut-off from families (communication bar) feeling of isolation and powerlessness too add to the stressors. Overloaded thus mentally and physically they face the risk to life and injuries which stands already amplified by living amidst lethal weapons of mass destruction.

2. The External Environmental Risk and Protective Factors

It includes the balance and interaction of fatal risks and protective factors. Military personnel live in a risk-prone situation, and, for them, it is not possible to avoid risk in a military setting. So we should focus on enhancing protective factors like resilience building, unit cohesiveness, hardy leadership, recruiting and deploying resilient soldiers, high level of preparedness, promote soldiers' collective efficacy, improved human resource practice within military team structure, timely positive and negative reinforcement, psychological debriefing, psychological first aid, and so on. A number of researchers (Bry, 1983; Magnusson, 1988; Rutter, 1993) have found that youth can adjust reasonably well to one or two risk factors or processes, but beyond two risk factors, the damage increases rapidly. In the military work settings and particularly during deployment in war, there is abundance of risk factors and, therefore, the quantum of probable damage is multiplicative. Research suggests that increasing the number of protective processes can help to buffer those risk mechanisms (Dunst, 1995; Rutter, 1993; Sameroff and Chandler, 1975). Increased deployments entail other stressful changes in military units as well, such as an increased number (and intensity) of training exercises, planning sessions and equipment inspections, all of which increase the workload and pace of operations (Castro and Adler, 1999). Furthermore, *more frequent deployments* also involve *family separations*, a recognized stressor for soldiers (Bell, Bartone, Bartone, Schumm, and Gade, 1997). Recently, Matthews (2008) observed that military as a perfect 'home' for positive psychology and Comprehensive Soldier Fitness (CSF) is a unique approach to behavioural health in a large organization.

3. Person–environment Interactional Processes

Whatever be the context/environment and the stressor, risks and protective factors are embedded within it, and a person (soldier) interacts by way of perceiving, selecting, modifying explicitly or implicitly some or all of the factors. In fact, it is the starting point of the interactional process where the individual matters and, therefore, it is assumed that resilience is teachable/trainable (Reivich and Shatte, 2002; Seligman, 1990; Reivich, 2010). Soldiers' ability can be enhanced by training them to know their strengths, understanding others and regulating their cognitions and emotions.

4. Internal Resiliency Factors

Internal personality or cognitive capabilities have been organized according to the prior-mentioned resilience framework for self-factors into five major cluster variables:

(i) Spirituality or Motivational Characteristics

Our cognitions and beliefs are intrinsic motivators that trigger and direct our efforts to face adversity. These may be individualistic or of higher levels like spirituality, meaning in life, optimism, etc. Believing in the purpose of life, inner drive to help others, conviction on the purpose of their pain and suffering, belief in the healing through helping/caring others are the predominant characteristics of resilient people (Beardslee, 1983, 1989; Segal, 1986; Werner, 1986). Taylor (1983) found that maintaining hope in adversity helps the individual to achieve environmental mastery and perceived control. Purpose in life or existential meaning (Frankel, 1959) helps resilient individuals to endure hardships, because they believe that they must survive to complete their mission. Purpose in life may be part of a larger latent construct called spirituality (Dunn, 1994). Religious faith and affiliation is an important factor of individual resilience (Masten, 1994). Independence and belief in oneself and one's uniqueness or speciality is an important trait of a person who tends to motivate resilient individuals towards positive achievement (Gordon and Song, 1994). Resilient individuals have more internal locus of control (Campbell, Converse and Rodgers, 1976; Luthar, 1991; Murphy and Moriarty, 1976; Werner and Smith, 1992) and are more hopeful about their ability to create positive outcomes for themselves and others. Perseverant (Bandura, 1989) and determined (Werner, 1986) cognitive style help the youth to be successful in chosen mission or direction.

Motivation and purpose in life can enhance through timely positive and negative reinforcement. Commanders can ensure a system of rewards and punishments contingent upon fair and objective evaluation of mission effectiveness. An excellent example of adequate and timely positive reinforcement is the incident when Major General David Tennent Cowan pinned his own Military Cross to wounded Field Marshal Sam Manekshaw (the then Captain) while fighting the battle against Japanese army during World War II. Identification and practicing of strength is a unique experience. Through this practice, a soldier identifies his strengths and feels one's uniqueness or specialty. It helps all to build resilience in military. Curiosity represents one's intrinsic desire for experience and knowledge. It involves the active recognition, pursuit, and regulation of one's experience in response to challenging opportunities and developed resilience.

(ii) Cognitive Competencies

Intelligence helps to buffer or reduce life stress (Masten *et al.*, 1988). According to Wolin and Wolin (1993), *insight* is the most important resiliency. It is the mental habit of asking penetrating questions of oneself and subsequently providing honest answers. Resilient youths have higher *self-esteem* associated with an accurate appraisal of their increased strengths and capabilities.

They have resilient *self-efficacy* (Bandura, 1977, 1989) and the ability to restore self-esteem (Flach, 1988) after failure or disruption in homeostasis.

Planning ability has been found related to resilience in high-risk youth (Anthony, 1987; Rutter and Quinton, 1984a). Researchers have reported that problem-solving ability is a component of resilience (Anthony, 1987; Neiger, 1991; Rutter and Quinton, 1984b). An ability to foresee consequences of choices and to plan a bright future is the characteristics of individuals who successfully overcome negative environments. Mann, Harmoni and Powers (1989) suggested that planned behaviour can be successfully taught. Jacobs and Wolin (1991) believe that *creativity* in a person allows them to express and resolve inner conflicts through painting, photography, dance, music and writing. A soldier/officer should be creative; through creativity he can find immediate practical solutions to problems on hand. Creativity is often seen as a sign of mental health and emotional well-being. Creativity in military organizations does matter not only in wartime, but also in peaceful times. Creativity counters stagnation and predictability and provides a competitive edge.

Judgement and *Critical thinking* are the cognitive competencies that help in facing challenges and building resilience. Emotional-motivational state of curiosity appears to fuel positive emotions such as excitement, enjoyment and attentiveness (Ainley, 1998; Kashdan and Roberts, 2002), facilitating complex decision making (Kreitler, Kreitler and Zigler, 1974) and goal perseverance (Sansone and Smith, 2000).

Cognitive competence: Academic and intellectual skill can be measured at the time of recruitment and deployment. Personal insight, self-esteem, planning ability and creativity can enhance through various cognitive tasks, puzzle solving (requires both physical and mental loads), and team/unit based activities. Taggar (2002) designated team creativity-relevant processes (for example, team citizenship, performance management, effective communication, involving others, providing feedback and conflict management) where the interactions among the team members support the application of individual creative resources and result in creative group performance.

(iii) Behavioural/Social Competencies

Citizenship, social responsibility, loyalty, teamwork, leadership, integrity, vitality and zest are the behavioural/social competencies. Citizenship represent a feeling of identification with and sense of obligation to a common good that includes the self, but that stretches beyond one's own self-interest. Youths from better-educated families are more positive about civic participation (Chapman, Nolin and Kline, 1997), and youths who get involved in political or quasi-political activities tend to be more tolerant (Avery, et al., 1992). Youths who are capable of acting competently in several cultures are more successful. Additionally, bi-gender competence is related to increased resilience and success in life among women (Dunn, 1994). Wolin and Wolin (1993) reported that resilient individuals have a capacity for intimacy. Many researchers (for example, Iversen for example, 2008; McTeague, McNally and Litz, 2004; Rona *et al.*, 2009) observed that relationship of improved well-being and readiness with higher levels of unit cohesion; and, unit cohesion which may serve as a resilience factor for Post-traumatic Stress Disorder (PTSD) and combat stress reactions.

Leadership is an interpersonal process, so we are focused on the psychological interaction between leaders and the led. Hardiness has emerged as a set of personal characteristics that help people turn stressful circumstances from potential disasters into opportunities for enhanced performance, leadership, conduct, health and psychological growth (Maddi, 2002; Maddi and Kobasa, 1984). Leaders who are high in hardiness themselves exert influence on their subordinates to interpret stressful experiences in ways characteristic of high-hardy persons. Several studies with cadets in training to be military officers lend support to the notion that leaders, high in hardiness, may influence subordinates to think and behave in more hardy or resilient ways (Bartone, 1999; Milan, Bourne, Zazanis and Bartone, 2002). Hardy leaders are exemplary in the way they handle the challenges by inspiring the entire social unit under their command. They are eager to take on the challenges as it further develops hardiness in them. Behavioural/social competencies like citizenship, teamwork, leadership can enhance through collective efficacy, multi-cultural experiences, interactive and good relation between troops soldiers and officers.

(iv) Emotional Stability and Emotional Management

Self-regulation has two faces: (1) cognitive regulation and (2) emotional self-regulation; both bring essential features of resilience at individual level. In the military context, there is always an intermingling of cognition, emotions and vigour. No training situation can replicate the real war situation. The successful outcome of the war always requires the global/total application of all such resources available in the situation. We can employ them effectively only when we are in a position to manage our emotions. It is beyond emotional intelligence. Some psychological skills like mental imagery, self-talk, goal-setting and organizational skills can be learned that help in managing emotional instability at the time of adversity (Eccles and Feltovich, 2008). Resilient individuals are characterized as reasonably happy people who are at least not prone to depression or negative appraisals of the reality characteristic of depressed individuals. Resilient people work well and play well (Garmezy, 1974). Resilient individuals recognize feelings and can control undesirable feelings such as fear, anger and depression.

Emotional stability/management: Selection and deployment of soldiers for the mission that need a higher level of resilience should be followed by an emotional intelligence test and development of emotional stress management skills.

(v) Physical Well-being and Physical Competencies

Good health (Werner and Smith, 1982, 1992), health maintenance skills (exercise, good diet and sleep), physical talent development (Masten, 1994) and physical attractiveness (Kaufman and Zigler, 1989) have demonstrated that good physical status is predictive of resiliency. *Vitality* describes a dynamic aspect of well-being marked by the subjective experience of energy and vitality (Ryan and Frederick, 1997). At the somatic level, vitality is linked to good physical health and body functioning, as well as freedom from fatigue and illness. On the psychological level, vitality reflects experiences of volition, effectance and integration of the self at both intrapersonal and interpersonal levels. Psychological tensions, conflicts and stressors detract

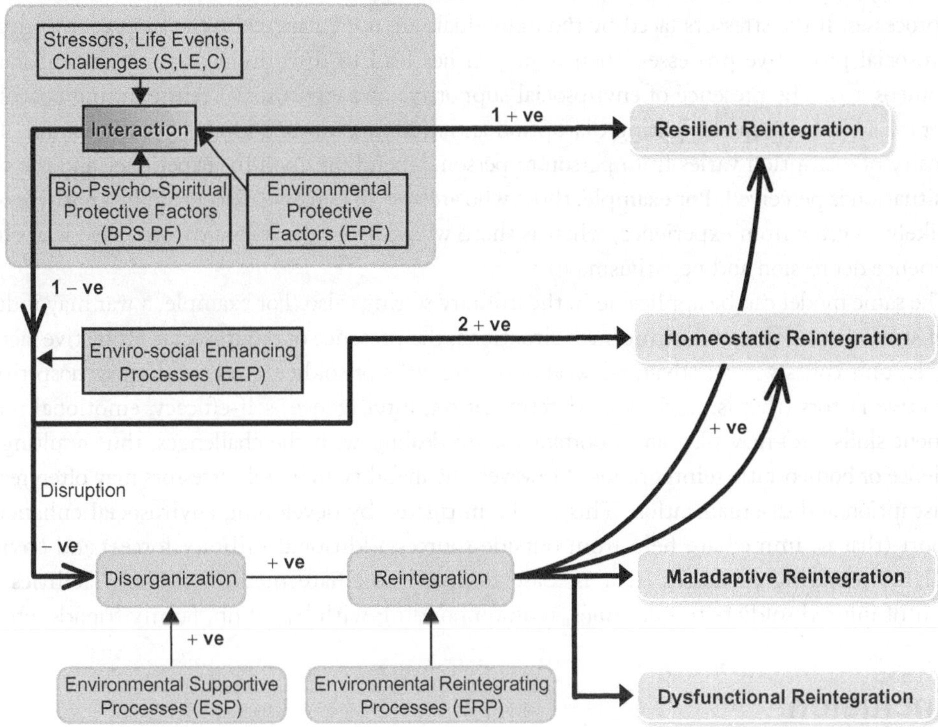

Figure 11.2 Adapted Version of Kumpfer (1999) Resilience Process Model.

from experienced vitality. At a strictly psychological level, vitality is associated with experiences of autonomy, effectance and relatedness. Vitality not only appears to be a correlate of health and well-being, but, also may actually contribute to it.

Physical competencies can enhance through autonomy, intimacy and relatedness. Support for autonomy has been shown to be positively associated with an individual's vitality (Ryan and Frederick, 1997). Autonomy and relatedness or intimacy increases vitality (Reis, Sheldon, Gable, Roscoe and Ryan, 2000). In the Indian military context, our selection, training as well as daily routine practices mostly focus on physical competencies of soldiers in spite of motivational, cognitive, behavioural and emotional competencies.

5. Resilience processes

Resiliency model shows that stressors or life challenges not balanced by external enviro-social protective processes or bio-psycho-spiritual resiliency factors in an individual can lead to imbalances in homeostasis or disruption (Flach, 1988). The resilience process model created by Kumpfer (1999) is multi-level (Figure 11.2).

The Resiliency Model is based on two aspects: (1) reintegration, and (2) intervention processes. There are four levels of reintegration: (1) resilient, (2) homeostatic, (3) maladaptive, and (4) dysfunctional reintegration. Likewise, there are four intervention processes, namely: (1) envirosocial protective, (2) envirosocial enhancing, (3) envirosocial supportive, and (4) envirosocial reintegrating processes. If the stressors faced by the individuals are not balanced by the biopsychospiritual/envirosocial protective processes, then it may either lead to disruption or result in imbalance of homeostasis. The presence of envirosocial supportive and envirosocial reintegrating processes may play a vital role in mitigating disruption and attaining various levels of reintegration. The intensity of disruption varies from person to person depending upon his experience and the way the situation is perceived. For example, those who are able to perceive disruption in a positive way are likely to grow from experience, whereas those who fail to take it in a constructive way often experience depression and negativism.

The same model can be applicable in the military settings also. For example, a war may induce lot of stressors on soldiers. During such situations, the presence of envirosocial protective factors (that is, unit cohesiveness, advanced weaponry, strengths of soldier, etc.) and biopsychospiritual protective factors (that is, faith, goal determination, intelligence, self-efficacy, emotional management skills, etc) may play an important role in dealing with the challenges, thus resulting in resilience or homeostatic reintegration. However, the inability to handle stressors may often result in disruption and disorganization. This can be mitigated by developing envirosocial enhancing support (that is, immediate help from outside sources/additional military forces) and envirosocial reintegrating processes (that is, base camps, immediate medical facility, facilities for transit of injured soldiers to safe zones, reunion bonding with battalion, family/friends, etc.).

Conclusion

It goes undoubtedly true that the operational situations and deployment conditions of soldiers always remain coupled with numerous inherent stressors and environmental pressures. Keeping these in mind, there is always a paramount requirement of fostering resilience among the military personnel so as to make them prepared to face the adverse situations boldly and recover from their negative influences in better and adaptive ways. It can be achieved by means of qualitatively enhancing and improving various types of protective factors around soldiers. Protective factors such as unit and family environment, social support system, welfare-oriented supports, and trustworthy leadership, etc., are some of the helpful factors in this pursuit. Qualitatively improved environmental protective factors will help generate 'positive person-environment interactions' which, in turn, will create an optimistic psychological climate around soldiers that will be conducive for focusing on their internal strengths while facing adversities. As it has been observed that resilience is trainable, impetus should be given to run dedicated programmes for resilience building among soldiers. Such programmes will certainly prove beneficial for the development of 'personal internal resiliency' factors among soldiers and, thereby making soldiers to appropriately use the various adaptive processes during adverse circumstances. Also, since there exists a relatively scarce empirical work in the domain of military resilience, more empirical researches concentrated on military population will make a long-lasting positive contribution towards making soldiers more resilient in future.

References

Adger, W. N. (2003). Social Capital, Collective Action and Adaptation to Climate Change: *Economic Geography, 79*(4), 287–404.

Ainley, M. D. (1998). Interest in Learning and the Disposition of Curiosity in Secondary Students: Investigating Process and Context. In L. Hoffman, A. Krapp, K. A. Renninger, and J. Baumert (Eds.), *Interest and Learning: Proceedings of the Seeon Conference on Interest and Gender* (pp. 257–66). Kiel, Germany: IPN—The Leibniz Institute for Science and Mathematics Education.

Anthony, E. J. (1987). Risk, Vulnerability and Resilience: An Overview. In E. J. Anthony and B. Cohler (Eds.), *The Invulnerable Child.* (pp. 3–48). New York: The Guilford Press.

American Psychological Association (2011). *The Road to Resilience.* Washington, DC: American Psychological Association. Retrieved from http://www.apa.org/helpcenter/road-resilience.aspx#

Avery, P., Bird, K., Johnstone, S., Sullivan, J., and Thalhammer, K. (1992). *Do All of the People Have All of the Rights All of the Time? Exploring Political Tolerance with Adolescents* (unpublished paper), Minneapolis: University of Minnesota, Center for Applied Research and Educational Improvement.

Bandura, A. (1977). Self Efficacy: Toward a Unifying Theory of Behavior Change. *Psychological Review, 84*(2), 191–215.

Bandura, A. (1989). Human Agency in Social Cognitive Theory. *American Psychologist, 44*(9), 1175–84.

Bartone, P. T. (1999). Hardiness Protects Against War-related Stress in Army Reserve Forces. *Consulting Psychology Journal, 51*(2), 72–82.

Beardslee, W. R. (1983). Commitment and endurance: A Study of Civil Rights Workers Who Stayed. *American Journal of Orthopsychiatry, 53*(1), 34–42.

Beardslee, W. R. (1989). The Role of Self-understanding in Resilient Individuals: The Development of a Perspective. *American Journal of Orthopsychiatry, 59*(2), 266–78.

Bell, D. B., Bartone, J., Bartone, P. T., Schumm, W. R., and Gade, P. A. (1997). *USAREUR Family Support During Operation Joint Endeavor: Summary Report* (ARI Special Rep. No. 34). Alexandria, VA: U.S. Army Research Institute for the Behavioral and Social Sciences.

Bry, B. H. (1983). Predicting Drug Abuse: Review and Reformulation. *The International Journal of the Addictions, 18*(2), 223–33.

Campbell, A., Converse, P. E., and Rodgers, W. L. (1976). *The Quality of American Life.* New York: Russell Sage Foundation.

Castro, C. A., and Adler, A. B. (1999). *Military Deployments and Soldier Readiness.* Proceedings of the 35th International Applied Military Psychology Symposium, Florence, Italy.

Chapman, C., Nolin, M., and Kline, K. (1997). *Student Interest in National News and Its Relation to School Courses (NCES 97–970).* Washington, DC: U.S. Department of Education, National Center for Education Statistics.

Delahaij, R., Gaillard, A. W. K., and Soeters, J. M. L. M. (2006). Stress Training and New Military Environment. In T. W. Britt, A. B. Adler, and C. A. Castro (Eds.), *Military Life: The Psychology of Serving in Peace and Combat. Vol. 2: Operational Stress* (pp. 141–50). Westport: Praeger.

Drury, J. (2009). Managing Crowds in Emergencies: Psychology for Business Continuity. *The Business Continuity Journal, 3*(3), 14–24.

Dunn, D. (1994). *Resilient Reintegration of Married Women with Dependent Children: Employed and Unemployed.* Doctoral Dissertation, Department of Health Education, University of Utah, Salt Lake City, Utah.

Dunst, C. (1995). *Risk and Opportunity Factors Influencing Child and Family Behavior and Development.* Presentation at 4th National Early Intervention Meeting. Coimbra, Portugal.

Eccles, D. W., and Feltovich, P. J. (2008). Implications of Domain-general "Psychological Support Skills" for the Transfer of Skill and the Acquisition of Expertise. *Performance Improvement Quarterly, 21*(1), 43–60.

Flach, F. F. (1988). *Resilience: Discovering New Strength at Times of Stress.* New York: Ballantine Books.

Frankel, V. E. (1959). *Man's Search for Meaning: An Introduction to Logotherapy.* Boston: Beacon Press.

Garmezy, N. (1974). Children at Risk: The Search for Antecedents of Schizophrenia. Part-1. Conceptual Models and Research Methods. *Schizophrenia Bulletin, 1*(8), 14–90.

Gorden, E. W., and Song, L. D. (1994). Variations in the Experience of Resilience. In M. C. Wang and E. W. Gordon (Eds.) *Educational Resilience in Inner-City America: Challenges and Prospects,* (pp. 27–43). Hillsdale, NJ: Erlbaum.

Iversen, A. C., Fear, N. T., Ehlers, A., Hacker, J., Hull, L., Earnshaw, M., Greenberg, N., Rona, R., Wessely, S., and Hotopf, M. (2008). Risk Factors for Post-traumatic Stress Disorder Among UK Armed Forces Personnel. *Psychological Medicine, 38*(4), 511–22.

Jacobs, J., and Wolin, S. (1991). *Resilient Children Growing Up in Alcoholic Families.* Paper Presented at National Consensus Symposium on Children of Alcoholics and Co-Dependence.

Kashdan, T. B., and Roberts, J. E. (2002). *Individual Differences in Interpersonally-generated Positive and Negative Affect.* Unpublished Manuscript, University at Buffalo, State University of New York.

Kaufman, J., and Zigler, E. (1989). The Intergenerational Transmission of Child Abuse. In D. Cicchetti and V. Carlson (Eds.), *Child Maltreatment: Theory and Research on the Causes and Consequences of Child Abuse and Neglect* (pp. 129–50). Cambridge: Cambridge University Press.

Kreitler, S., Kreitler, H., and Zigler, E. (1974). Cognitive Orientation and Curiosity. *British Journal of Psychology, 65*(1), 43–52.

Kumpfer, K. L. (1999). Factors and Processes Contributing to Resilience. In M. D. Glantz and J. L. Johnson (Eds.), *Resilience and Development: Positive Life Adaptations* (pp. 179–224). New York: Kluwer Academic.

Lentzos, F., and Rose, N. (2009). Governing Insecurity: Contingency Planning, Protection, Resistance. *Economy and Society*, *38*(2), 230–54.

Luthar, S. (1991). Vulnerability and Resilience: A Study of High-risk Adolescents. *Child Development, 62*(3), 600–16.

Maddi, S. R. (2002). The story of Hardiness: 20 Years of Theorizing, Research and Practice. *Consulting Psychology Journal, 54*(3), 173–85.

Maddi, S. R., and Kobasa, S. C. (1984). *The Hardy Executive: Health Under Stress.* Homewood, IL: Dow Jones-Irwin.

Magnusson, D. (1988). *Individual Development from an Interactional Perspective.* Hillsdale, NJ: Erlbaum.

Mann, L., Harmoni, R., and Power, C. (1989). Adolescent Decision-making: The Development of Competence. *Journal of Adolescence, 12*(3), 265–78.

Masten, A. S. (1994). Resilience in Individual Development: Successful Adaptation Despite Risk and Adversity. In M. C. Wang and E. W. Gorden (Eds.), *Educational Resilience in Inner-city America* (pp. 3–25). Hillsdale, NJ: Erlbaum.

Masten, A. S. (2001). Ordinary Magic: Resilience Processes in Development. *American Psychologist, 56*(3), 227–38.

Masten, A. S., Garmezy, N., Tellegen, A., Pellegrini, D. S., Larkin, K., and Larsen, A. (1988). Competence and Stress in School Children: The Moderating Effects of Individual and Family Qualities. *Journal of Child Psychology and Psychiatry and Allied Disciplines*, 29(6), 745–64.

Matthews, M. D. (2008). Positive Psychology: Adaptation, Leadership, and Performance in Exceptional Circumstances. In P. A. Hancock and J. L. Szalma (Eds.), *Stress and Performance* (pp. 163–80). London: Ashgate.

Mayunga, J. S. (2007). *Understanding and Applying the Concept of Community Disaster Resilience: A Capital Based Approach.* A Working Paper Prepared for the Summer Academy for Social Vulnerability and Resilience Building, 22–28 July 2007, Munich, Germany.

McAslan, A. (2010). *The Concept of Resilience: Understanding Its Origins, Meaning and Utility.* Adelaide: Torrens Resilience Institute.

McTeague, L., McNally, R., and Litz, B. (2004). Prewar, War-zone, and Postwar Predictors of Post-traumatic Stress in Female Vietnam Veteran Health Care Providers. *Military Psychology, 16*(2), 99–114.

Milan, L. M., Bourne, D. M., Zazanis, M. M., and Bartone, P. T. (2002). *Measures Collected on the USMA Class of 1998 As Part of the Baseline Officer Longitudinal Data Set (BOLDS)* (ARI Tech. Rep. No. 1127). Alexandria, VA: U.S. Army Research Institute for the Behavioral and Social Sciences.

Murphy, L. B., and Moriarty, A. E. (1976). *Vulnerability, Coping and Growth: From Infancy to Adolescence.* New Haven, CT: Yale University Press.

Neiger, B. (1991). *Resilient Reintegration: Use of Structural Equations Modeling.* Unpublished Doctoral Dissertation, University of Utah, Salt Lake City.

Norris, F. H., Stevens, S. P., Pfefferbaum, B., Wyche, K. F., and Pfefferbaum, R. L. (2007). Community Resilience as a Metaphor, Theory, set of Capabilities and Strategy for Disaster Readiness. *American Journal of Community Psychology, 41*(1–2), 127–50.

Norris, F. H., Tracy, M., and Galea, S. (2009). Looking for Resilience: Understanding the Longitudinal Trajectories of Responses to Stress. *Social Science and Medicine*, *68*(12), 2190–98.

Paton, D., and Johnston, D. (2006). *Disaster Resilience: An Integrated Approach.* Springfield: Charles C. Thomas Publications Ltd.

Pendall, R., Foster, K. A., and Cowell, M. (2007). *Resilience and Regions: Building Understanding of the Metaphor.* A Working Paper for Building Resilience Networks. Institute of Urban Regional Development, University of California.

Reis, H. T., Sheldon, K. M., Gable, S. L., Roscoe, J., and Ryan, R. M. (2000). Daily Well-being: The Role of Autonomy, Competence, and Relatedness. *Personality and Social Psychology Bulletin, 26*(4), 419–35.

Reivich, K. J. (2010). *Master Resilience Training Trainer Manual.* Unpublished Manual, University of Pennsylvania, Philadelphia, U.S.

Reivich, K., and Shatte, A. (2002). *The Factor: Seven Essential Skills for Overcoming Life's Inevitable Obstacles.* New York, NY: Broadway Books.

Rona, R. J., Hooper, R., Jones, M., Iversen, A. C., Hull, L., Murphy, D., Hotopf, M., and Wessely, S. (2009). The Contribution of Prior Psychological Symptoms and Combat Exposure to Post Iraq Deployment Mental Health in the UK military. *Journal of Trauma Stress, 22*(1), 11–19.

Rutter, M., and Quinton, D. (1984a). Parental Psychiatric Disorder: Effects on Children. *Psychological Medicine, 14*(4), 853–80.

Rutter, M., and Quinton, D. (1984b). Long-term Follow-up of Women Institutionalized in Childhood: Factors Promoting Good Functioning in Adult Life. *British Journal of Developmental Psychology, 2*(3), 191–204.

Rutter, M. (1993). Resilience: Some Conceptual Considerations. *Journal of Adolescent Health*, 14(8), 626–31.

Ryan, R. M., and Frederick, C. (1997). On Energy, Personality, and Health: Subjective Vitality as a Dynamic Reflection of Well-being. *Journal of Personality, 65*(3), 529–65.

Sameroff, A. J., and Chandler, M. J. (1975). Reproductive Risk and the Continuum of Caretaking Casualty. In F. D. Horawitz (Ed.), *Review of Child Development Research* (pp.187–244). Chicago: University of Chicago Press.

Sansone, C., and Smith, J. L. (2000). Interest and Self-regulation: The Relation between Having To and Wanting To. In C. Sansone and J. M. Harackiewicz (Eds.). *Intrinsic and Extrinsic Motivation: The Search for Optimal Motivation and Performance* (pp. 341–72). San Diego, CA: Academic Press.

Segal, J. (1986). *Winning Life's Toughest Battles: Roots of Human Resilience.* New York: McGraw Hill.

Seligman, M. E. P. (1990). *Learned Optimism.* New York, NY: Knopf.

Seville, E., Brunsdon, D., Dantas, A., Le Masurier, J., Wilikinson, S., and Vargo, J., (2006). *Building Organizational Resilience: A Summary of Key Research Findings.* Christchurch, New Zealand: Resilient Organizations Research Programme.

Sharma, S., and Sharma, M. (2008). Stress-induced Psychological Vital Signs in Military Domain: Assessment and Intervention Issues. *Psychological Studies, 53*(1), 7–19.

Stetz, M. C., Thomas, M. L., Russo, M. B., Stetz, T. A., Wildzunas, R. M., McDonald, J. J., Wiederhold, B. K., and Romano, J. A. Jr. (2007). Stress, Mental Health and Cognition: A Brief Review of Relationships and Countermeasures. *Aviation, Space, and Environmental Medicine, 78*(5), 252–60.

Taggar, S. (2002). Individual Creativity and Group Ability to Utilize Individual Creativity Resources: A Multilevel Model. *Academy of Management Journal, 45*(2), 315–30.

Taylor, S. E. (1983). Adjustment to Threatening Events: A Theory of Cognitive Adaptation. *American Psychologist, 38*(11), 1161–73.

Werner, E. E. (1986). Resilient Offspring of Alcoholics: A Longitudinal Study from Birth to Age 18. *Journal of Studies on Alcohol*, 47(1), 34–40.

Werner, E. E., and Smith, R. S. (1982). *Vulnerable But Invincible: A Longitudinal Study of Resilient Children and Youth.* New York: McGraw-Hill Book Co.

Werner, E. E., and Smith, R. S. (1992). *Overcoming the Odds: High Risk Children from Birth to Adulthood.* Ithaca, NY: Cornell University Press.

Wolin, S. J., and Wolin, S. (1993). *Bound and Determined: Growing up Resilient in a Troubled Family.* New York: Villard Press.

Chapter 12

Stress and Resilience Capacity of Students: A Burning Public Health Issue

Sibnath Deb, Aneesh Kumar and Anjali Gireesan

Department of Applied Psychology, Pondicherry University, Puducherry

Abstract Stress among students has become a global public health issue because of over expectation of parents, academic competition and limited job opportunities. Stress has different connotations in different contexts of our lives. The present secondary review-based study attempts to highlight the stress experienced by students at various levels, causal and contributing factors behind the same and its consequences. The chapter also throws light on the resilience capacity of the students and its role in managing stressful events. Evidences indicate that the most common stress experienced by the students is academic stress resulting from over expectation and academic workload followed by stress caused by social pressure and peer comparisons. The other types of stress include poor peer acceptance, disturbed family environment, conflicts in interpersonal relations, problems and anxiety related to transition and changes in life. The fallout of academic stress is gradual deterioration of physical and mental health including cognitive and affective functioning. It has also been observed that the management of stress by students depends on their resilience capacity. A slight variation in performances even in case of high achievers can cause internal stress which may be challenging to cope up with and would even take to the extent of suicidal ideation and attempt. Intervention programmes indicate that stress can be managed though various mind–body programmes. In fact, it might be stated that students experience a range of stress on myriad issues and these can be managed through enhancing their resilience capacity. It can be inferred from the studies that there is an urgent need to take up institution-based intervention programme to empower students with knowledge and information about various issues which disturb adolescents and also enhance their resilience capacity to face challenges in life. Again, this calls for recruitment of student counsellors or psychologists, in all educational institutions, for addressing mental health needs of the students.

The importance of mental health as a key component in child and adolescent development is beginning to influence global health initiatives (Sawyer, Afifi, Bearinger, Blakemore and Dick, 2012). This is of particular significance because it is predicted that when the current generation of children become adults, depression will be the second greatest contributor of disease, globally (Mathers and Loncar, 2006). Stress is a subjective sensation with a varied degree of perception. Stress includes a wide range of strong external stimuli, both physiological and psychological, which can cause a physiological response called, 'the general adaptation syndrome', first described in 1936 by Hans Selye in the journal *Nature*. The transition from *eustress* (curative stress or positive form of stress) to distress (having negative implications) occurs when the demands exceed the personal and social resources that the individual is able to mobilize (Selye, 1956). Distress is generally triggered by a disturbing/unexpected event in one's life, for example, failure in exams, divorce, notification of debt, tax, etc. It affects an individual's mental and physical health negatively.

Stress relates to all minor and major events in one's life. It has become a matter of pubic heath and concern (Shahmohammadi, 2011). Stress among students is a major concern as the future and development of a country lies in the hands of the youth. It would affect a child's mental, physical and social functioning. Although the literature tends to focus on the impact of single-variable stressors on children's development, in real-life situations, children experience stress from multiple sources (Jewett and Peterson, 2010). Experts suggest that globally children experience stress and they manifest it through depression or avoidance (Jewett and Peterson, 2002). Among children, stress is most often observed as an overt physical reaction which, generally, is characterized by behaviours like crying, running away from home, emotional outbursts, and self-harming behaviours, headaches and stomach aches, or even motor behaviours (for example, hair twirling, thumb sucking, biting fingernails), toileting emergencies and sleep disturbances (Jewett and Peterson, 2010).

Models of Stress

The term *stress* has been explained, mainly with the help of medical, environment and psychological models (McNamara, 2000; Suldo, Shaunessy and Hardesty, 2008). In the medical model, stress has been explained as a state of distress in an individual in response to an environmental antecedent. This change in physiological response of an individual can be measured by increased heart rate, elevated blood pressure and the presence of hormones and neurotransmitters like cortisol, adrenaline and norepinephrine that heighten the arousal sates within an organism (Selye, 1956). Distress experienced within a normal range, is adaptive in nature; this heightened arousal prepares an organism to effectively deal with stress. However, in the long term, chronic stress which is a consequence of prolonged distress has been linked to poor health conditions including both physical (Raguram, Weiss, Keval and Channabasavanna, 2001; Sundblad, Jansson, Saartok, Renström and Engström, 2008) and mental well-being (Wang and Ollendick, 2001). In the environmental model, stress is defined as external factor, including threats and environmental problems. External stress can cause negative outcomes like anxiety, depression and aggression; also it can affect academic achievement, cause substance abuse and poor life satisfaction. Although biological and environmental stress have both been linked to negative outcomes of stress for adolescents, neither of the models sufficiently explains why some adolescents who experience similar types of stress do not experience the negative outcomes while in some others it was manifested (Suldo *et al.*, 2008).

Psychological model of stress focuses on the concept of perceived stress, which refers to interactions between an environmental precipitant (external stress); the physiological reactions of the body (distress); and a person's cognitive, emotional and behavioural response to this interaction (ibid.). Stress is perceived when an external event causes aversive physiological and cognitive distress in an individual that exceeds his or her emotional and behavioural repertoire designed to negate the harmful effects of external stressors. The conceptualization of perceived stress allows for consideration that certain individuals may possess resources, such as coping that allow them to experience external stress without experiencing compromised functioning (Luthar, Sawyer and Brown, 2006). In recent years, this transactional perspective of stress has come to be regarded as the most widely accepted and cited definition of stress. Children and adolescents who report high levels of perceived stress are at high risk for negative outcomes, such as depression, substance abuse (Sinha and Jastreboff, 2013), academic underachievement (Deb and Walsh, 2011; Schmeelk–Cone and Zimmerman, 2003) and diminished life satisfaction (Shahmo–hammadi, 2011).

Causes of Stress

Researchers observed that multiple stressors interact with one another and can have cumulative effects (Jewett and Peterson, 2010). Being physical or emotional, internally or externally generated, the stressors are classified as external and internal combination of endogenous and exogenous factors and system-induced stress (Shahmohammadi, 2011). Exogenous or external factors are those factors which originate from outside and individual have no control over that. These external factors place demands or exert stress on the individual, which may include pressures and demands from people or situations. These factors may originate and can be from people known or unknown to the individual and, also from life events such as bereavements, illness, job and others. The stress among children due to external stressors would include separation from family, exposure to arguing and interpersonal conflict, exposure to violence, abuse or neglect.

Endogenous or internal factors include strong memories of previous unpleasant experiences, which make individuals highly sensitive to such instances. They anticipate the same unpleasant events, and their fear gives rise to feelings of inferiority, inadequacy and uncertainty (ibid.). Also internal stress would include somatic complaints, pain, hunger, etc. Some conditions including disability or chronic illness like cancer or HIV/AIDS can exert stress on individuals. The combination of endogenous and exogenous factors also leads to the creation of stress. In children, the experiences of rejection, failure, exposure to violence or neglect and related emotional or physical pain can later cause stress. This can affect development and adjustment in later stages of life. System-induced stress is another type of stress identified by psychologists where the system refers to organizations like school, family, office and other such social structures in which an individual functions (ibid.). Life becomes tough if any of these systems act as stress inducing. This type of stress is highly relevant for developing societies on the context of changing demands in the family and workplace. One group of society, which is a victim of system-induced stress, is the student community. The academic pressure and workload together with pressure, punishment and high expectations from parents and teachers, create burden and makes student life stressful and depressing (Kumar, Gireesan, and Deb, *In Press*).

Negative events, crises and challenges in life are unavoidable, evoking a plethora of human responses. Some individuals adapt while others fail to cope up with stressful experiences. Maintaining personal well-being requires skills and resources to overcome negative events. Resilience research explores protective factors and processes that result in positive outcomes despite elevated risk for maladjustment (Luthar, Sawyer and Brown, 2006; Masten, 2004). The broad objective of this chapter is to highlight the stress experienced by students at various levels, its casual and contributing factors and its consequences. This review-based chapter has generated evidence from national and international journals, periodicals and various government documents.

Students Stress: An Indian Scenario

The period from childhood to adolescence has varied changes and developments which are marked by physical and psychological changes. These changes and transition can also cause stress and pressure among them. But, rather than the stress caused by the sudden and abrupt developmental changes, the stress caused by family, academic and other environmental stressors is more challenging and affects the development and health of an individual (Howard and Medway, 2004). When we look back to Mahatma Gandhi's 'basic education' that was based on the cultural tenets of the Indian society with the introduction of a productive handicraft, related skills and knowledge systems in the school curriculum we find failed to make a long-term impact on the existing educational system in India. Under British influence, education in India became textbook-oriented, both for the teacher and the students, demanding that students memorize class material for reproduction in examinations. Being thorough in covering the content of textbook content was seen as a way of preparing the students for major examinations and for a secure life afterwards (Verma, Sharma and Larson, 2002). This has developed into the present education system.

Corporal punishment is a part of Indian culture where parents believe that punishing children is the best way to have control over them and/or to modify their behaviour. However, there is some evidence for corporal punishment in the ancient Indian history. There is an ancient dictum (*Chanakya-Neeti)* in the Indian history, *'Lalayet Pancha barsani, dasabarsani tarayat, prapte tu shodhase barse putra mitra badacharet'*, which means 'indulge the child up to five years, punish him up to ten years, and when he reaches sixteen, treat him as a friend' (Raj, 2011). Corporal punishment can be emotionally harmful to children. Researches indicate that children receive messages confusing love with pain, and anger with submission, which is psychologically harmful (Deb and Modak, 2010; Deb and Walsh, 2011). Many went on to argue that children cannot be disciplined without punishment. Studies have also found that children tend to admit that parents and teachers have the right to punish them and should never revert back, but these experiences tend to be stressful and painful for them (Deb, Kumar, Bhattacharyya and Jiandong, 2012; Kumar *et al., In Press*).

Stressors for children are either present at home or at school where they spend a substantial part of their childhood. These two places are often considered to be the safest and ideal places for their overall developments. At home, children experience corporal punishment in the hands of their parents. They often encounter abuse including physical, sexual or of emotional nature by their relatives and others, witness violence and drug abuse, and become victims of sibling rivalry or cultural beliefs. On the other hand, at school children become victims of bullying or experience punishment and poor relations with teachers in addition to exam anxiety, school workload, academic pressure and completion.

Pre-school Children and Stress

Childhood depression is widely under recognized in primary health care settings. This phenomenon appears to increase with younger age. Evidence has been provided for a valid depressive syndrome among pre-school children (Luby *et al.*, 2002, 2003). Based on the need for the earliest possible identification of depression, the development of a brief screening measure to capture young children with markers of depression from these community settings was developed and tested (Luby, Heffelfinger, Koenig–McNaught, Brown and Spitznagel, 2004). Most of the pre-school children do display typical symptoms and signs of stress and depression (Luby *et al.*, 2004). Stress in overall prevalence was consistent with rates for older children and was significantly related to minority status, male sex, low socioeconomic status, father's absence and small family size (Lavigne *et al.*, 1996).

Research evidence also suggests that preschool children can manifest typical symptoms of depression when age-adjusted symptoms states are assessed (Luby *et al.*, 2002). Anhedonia appeared to be a specific symptom and sadness/irritability appeared to be a sensitive symptom of preschool children with depression (Luby *et al.*, 2003). There is a significant need for early identification of emotional/behavioural problems in very young children. Stress and disruptions in family life may further contribute to risk for persistent problems in later stages (Briggs–Gowan, Carter, Skuban and McCue, 2001).

Stress among School Students

Indian children, today, are under severe stress. When we analyse a regular school student's life, we find they rush to school early in the morning and from there to tuitions, and then to coaching institutes. Some children even have more than two tutors and their classes start form early morning, even on Sundays and holidays. And finally they return home with sacks of assignments and projects. At home, parents pressurize them for better academic performance and for reaching their expectations in the process cutting down on their play time and leisure. Adding to the stress is the punishment children receive at home. School life is not much different. Teachers and tutors bombard children with workloads and tests. Schools have become factories which concentrate more on quantity and less on quality. They use corporal punishments and do little or nothing for the well-being of children; rather they concentrate more on marks received by the students which would uplift the school's popularity.

Stress among students in India can be generally classified into two major classes of stress related to academic work incorporating pressure, burnout, test anxiety, workload and stress due to other factors including relationship issues, violence, conflicts, and so on (Abraham, Zulkifli, Fan, Xin, and Lim, 2009). Children's psychological adjustment to the stress has been shown to affect educational progress which is critical for their future (Cortina *et al.*, 2013). The problem of test anxiety is much more significant in Indian and other Asian children as compared to non-Asian children (Wang and Ollendick, 2001). Indian students express a collectivistic nature and, also relate strongly to relationships with parents and elders, hence conflicting relations can be stressful for them (Verma *et al.*, 2002). A number of studies investigating the nature of anxiety and depression in Asian cultures in general, and in India in particular, have reported that Indian children tend to express psychological conflicts through somatic symptoms (Raguram *et al.*, 2001).

Academic Stress, Exam Anxiety and Workload

Indian classrooms have been termed as 'joyless' and loaded with academic burden by the National Advisory Committee report constituted in March 1992 by the Ministry of Human Resource Development (Verma *et al.*, 2002). In the current education system in India, students have to take three to four examinations per academic year. Particularly, the examinations at the end of the 10th and 12th grade have a significant impact on the lives of children and their parents (Shahmohammadi, 2011). Exam or test anxiety has been observed to be very common in children across the globe (Bodas and Ollendick, 2005). In India, anxiety and stress, due to academic pressure, is experienced among school students and adolescents due to high expectations and pressure by parents (Deb and Walsh, 2010; Deb *et al.*, 2012). It was reported that in a year in India alone 3130 children, approximately more than six children per day, committed suicide of which a major reason was failure in examinations and family issues. Also 7% suicides in Chandigarh were due to failure in examination (National Crime Records Bureau, 2010).

Performance in these exams is crucial for career formation including choosing of higher education at universities and colleges and also bringing prestige to the family (Bodas and

Ollendick, 2005). Such practices exert tremendous pressure on the students. It was found that the experience of being evaluated, critiqued or judged commonly on the basis of tests or exams resulted in emotional disturbances, uneasiness and negative feelings towards self and environment (Donaldson, Gooler and Scriven, 2002). Students' perceptions of their educational environment and their personality characteristics are linked with test anxiety (Bagana, Raciu, and Lupu, 2011). It has been found that test anxiety could affect a child's adjustment at home, school, social and emotional functioning (Bodas and Ollendick, 2005). With regard to the interactional effects of age, gender and educational level, Indian school-going females reported more test anxiety than those in college; whereas, Indian college males had higher test anxiety than their school-going counterparts (ibid.). A negative relationship is generally seen between self-esteem and optimism and test anxiety (Bagana *et al.*, 2011). The 'do or die' competitive educational system was thought to induce high levels of stress in students as well as parents.

School workload and experiences are also related to stress levels (de Anda *et al.*, 2000). Specifically, adolescents in India were found to spend majority of their quality time in school-related activities. Moreover, it was found that schoolwork generated negative emotional states including frustration and unhappiness. School workload including homework and tightly packed time table were found to be the major reasons to create pressure and stress among students (Shahmohammadi, 2011).

Peer Relations, Acceptance and Bullying

There is strong evidence that peer victimization including bullying, poor peer relations and experience of repeated harassment from peers, is detrimental to students' psychosocial development and is a major stressor (Troop–Gordon and Gerardy, 2012; Bauman, Toomey and Walker, 2013). Bullying and peer victimization is associated with increased risk for suicide among youth (Kim, Leventhal, Koh and Boyce, 2009). Both perpetrators and targets of peer bullying are at higher risk of suffering from depression, suicide ideation and are more prone to attempt suicide than adolescents not involved in bullying (Kim *et al.*, 2009; Bauman *et al.*, 2013). Poor peer relations in schools would negatively affect a student's mental health and it would result in poor scholastic interest and performances added with poor social relations and self-esteem (Kshirsagar, Agarwal and Bavdekar, 2007; Troop–Gordon and Gerardy, 2012). Females are more prone to internalizing and males to externalizing negative experiences (Bauman *et al.*, 2013).

Relation with Teachers

Majority of Indian teachers use corporal punishment for better academic performances and disciplining and express their dissatisfaction on academic performance by the child which in turn, leads to stress and other emotional problems even sometimes causing suicides (Deb *et al.*, 2012). Sometimes, a teacher's acts, attitudes and classroom preferences can have negative effect on the students (Sava, 2002). Most of the research studies focus on positive teacher communication behaviours, whereas fewer studies have been done on negative teacher communication behaviours which can really be detrimental to a student's development (Sava, 2002). Maltreatment, abuse and lack of support by teachers would have strong reflection on a student's academic, emotional and personal development (Newberry and Davis, 2008). Teachers tend to have a special liking to students, who stand out in their academic performance (Hughes, Zhang and Hill, 2006; Newberry and Davis, 2008), male students are liked by female teachers (Sava, 2002) and vice versa, and a particular dislike towards some students. Poor relations with teachers or children who are neglected in classrooms by teachers would develop psychosomatic symptoms, poor psycho-social development and low academic performances (ibid.) and the relationship with teachers can have sustained and greater impact on student's later life (Uitto, 2012). Children who are sexually abused by teachers tend to develop greater psychological problems of depression and suicidal ideation. Students, who have close relations with teachers have better peer relations, enjoy school and academic activities and also experience less stress at school (Hughes *et al.*, 2006).

Stress at Home: Sibling Rivalry

Children are at high risk for being victims of aggression and violence by their own siblings (Linares, 2006; Khan and Cooke, 2008). Exposure to sibling violence and rivalry would later become risk factors for impaired psychological functioning of the child which could be as internalizing symptoms like depression, anxiety and low-self esteem or externalizing symptoms like destructive behaviour and poor social interactions (Linares, 2006). More conflict and less warmth between siblings have been linked to problems including depression, anxiety and anti-social behaviours including alcohol and drug abuse (Khan and Cooke, 2008). Also, poor sibling relations lead to disruptions in school performances and interest (Linares, 2006). Sibling rivalry and violence may end up in teasing, ridiculing, insulting, threatening to even terrorizing or physical or sexual abuse including slapping, hitting, biting, kicking, touching or fondling, indecent exposure or to the extent of forced sexual acts (Cyr, Wright, McDuff, and Perron, 2002; Hardy, 2001). The stress experienced by sibling violence affects a child's mental and social functioning. It gives rise to high internal stress which is similar to characteristic features of post-traumatic stress disorder in later stages of life (Linares, 2006).

Family Environment and Parental Relations

Children who are brought up in loving, supportive and caring environments are more likely to behave the same way in later stages of life. Children who are brought up in fearful, anger and hostile environment have a higher chance of experiencing similar environments in adulthood. Parental symptoms of depression, family cohesion and family expressiveness emerged as significant predictors of child-reported distress (Jobe-Shields *et al.*, 2009). Parental relation affects a child's development in all levels including cognitive, mental and physical well-being and also in social relation with peers, school and others (Child Welfare Information Gateway, 2009). Children raised in different family contexts display differential behavioural and developmental patterns. In particular, children raised in single-parent families have been found to do less well across a range of measures of well-being than others. Also parental separation or divorce has been found to be associated with an array of adverse outcomes for children (Mackay, 2005; Pickhardt, 2009). Parent's remarriage and relations with step parents are also found to be risk factors to stress among children due to less affection, abuse and neglect (Pickhardt, 2009). Intra-parental conflict, partner violence including verbal or physical conflict would negatively affect the child's mental state leading to depression or even developing behavioural problems (de Anda *et al.*, 2000). Financial crisis, socio-economic conditions (Bodas and Ollendick, 2005) and drug or substance abuse (Child Welfare Information Gateway, 2009) among parents can greatly act as stressor for students. Children with poor family conditions and attachment with parents are found to have low self-esteem and socializing skills (ibid.).

Indian parents take active involvement in academic activities of their children. They emphasize on academic achievement as they control the child's leisure, play and social activities and often send their children to tuitions in order to advance their academic achievements even when their children perform well at school (Verma *et al.*, 2002; Deb *et al.*, 2012). In particular, feelings of shame and disgrace to oneself and one's parents and family may be associated with examination failure (Wang and Ollendick, 2001). Studies have also reported that parents' educational and occupational status was positively correlated with test anxiety and academic pressure in children (Bodas and Ollendick, 2005). A majority of Indian parents apply corporal punishment for better academic performances and for disciplining their children (Deb *et al.*, 2012). Corporal punishment and emotional neglect have been found to cause stress and anxiety among students (Muhisina, 2013). Lack of parental support and rejection would also be a causal factor to stress (Deb *et al.*, 2012).

Stress among College and University Students

College students also experience stress, but the stressors and its manifestations are different from that of children. Relationship problems, loveless and neglectful bonding with parents, physical abuse were found to be stressors for college students (Singh, Manjula and Philip, 2012). Also, academic burden, career and marriage conflict, transition from school to college and then to workplace, competition and exam stress, affect an adolescent's mental and physical well-being (Al-Sowygh, Alfadley, Al-Saif, and Al-Wadei, 2013). It was reported that in a year alone in India, family issues, failure in examinations and unsuccessful romantic relationships were reported as the major reasons to have driven 47,625 youths (aged between 15–29 years) to commit suicide (National Crime Records Bureau, 2010). Adverse childhood experiences can also manifest as depression in later stages (Singh *et al.*, 2012).

Transition from School to College

The students' transition from one to another level of their formal education can be experienced as challenging and stressful in life (Negovan and Bagana, 2011) because in these new situations individuals experience many significant and different events, where they find themselves in different and unfamiliar situations having to cope with new and sometimes difficult experiences (Jindal–Snape and Miller, 2008). The transition from high school to university or college education can also be considered as a crisis which also coincides with the transition from late adolescence to early adulthood. In the transition stage of their academic life, students experience three major stressful situations which may significantly affect their mental and physical well-being; the stressors being exposure to a new environment, being required to make new responses and the establishment or loss of significant relationships (Negovan and Bagana, 2011).

Changing Relationships

Several factors affect relationships and forms of relationships among adolescents (Tagay and Karataş, 2012). In Indian culture, romantic relationships are not much supported and students are expected to adhere to parental decisions. These cultural and societal norms put great pressure on adolescent relations and put them in a greater state of confusion with regard to decision making (Tagay and Karataş, 2012). Romantic relationships and the style of attachment it evokes have a great influence on the adolescent. Poor, insecure, and avoidant type of attachment style with partner results in stress and confusion in both partners. Relationship satisfaction, in general, refers to feelings, thoughts, or behaviours within a relationship associated with sexual attitudes, professed feelings of love, commitment, self-disclosure and relationship investment (Demir, 2008). Several studies have shown that the quality of romantic relationships is a significant source of happiness and subjective well-being (Demir, 2008; Demirtas and Tezer, 2012). Problematic romantic relations or failure in romantic relations would be stressful and may lead to depression (Demirtas and Tezer, 2012) and suicide (National Crime Records Bureau, 2010). Another major problem which adolescents face is lack of parental support, poor attachment with parents and neglectful parenting (Singh *et al.*, 2012). Relations with parents reflect in their personality and, also in all realms including academic performances and future aspirations (Anders, 2011). Poor family conditions both psychologically and economically would cause stress in students and also exert pressure on them (Byrne, Davenport and Mazanov, 2007).

Academic Burden and Workload

Assigned workload and performance pressure have been observed as the main perceived stressors among students who take up regular college or university education (Abraham *et al.*, 2009). Examinations, fear of failing, workload and completing course requirements ranked highest among stressors related to student in training and the academic environment (Byrne *et al.*, 2007; Trifoni and Shahini, 2011). It was found that students doing professional courses like medical and dental course experienced high workload, stress and pressure (Abraham *et al.*, 2009; Al-Sowygh *et al.*, 2013). They experienced stressful transition and exam anxiety when they passed from one year to another till the final year of the course. Lack of language skill was experienced as stressful by first year students. Inappropriate and conflicting evaluations and criticisms, and approachability to faculty were other reasons for stress (Trifoni and Shahini, 2011).

Career Planning and Marriage

In our society, boys are still considered to be the ones to undertake the economic responsibility of the family in future and are expected to obey the rules more in order to meet the expectations of the family and society. In view of these high expectations, men are said to regard the needs of others more than their own. On the other hand, girls are brought up as being scared of close relationships in our society and their submissive and obedient behaviour is supported. Marriage and career planning are also challenges for adolescents (Byrne *et al*., 2007). Adolescents who delay career planning and marriage tend to experience less conflict and stress (Barnett, Gareis, James and Steele, 2003).

Indian Cultural Beliefs, Society and Environmental Factors as Stressors

The cultural beliefs and practices of people relating to health is part of a society in which they live. The effect of culture is observed in all areas, including health and child care. Practices are often based on hearsay and can be detrimental to general health. Children represent the core group most affected by traditional practices. Child marriage and gender discrimination is also a great barrier for the well-being of children (Malhotra, Warner, McGonagle and Lee-Rife, 2011). A female child is considered as a burden for the family, where she is subjected to abuse and her needs including education and well-being are neglected (Gupta, 2007). Also, in some sections of the society education of children is not considered important and children are sent as child labourers or for forced begging (Ministry of Women and Child Development, 2007). Children with disability or mental illness are also at risk of victimization due to faulty beliefs (Edwardraj, Mumtaj, Prasad, Kuruvilla and Jacob, 2010). Children with chronic illness like HIV/AIDS and cancer also experience rejection from the society. There are many other cultural beliefs and societal attitudes which trigger stress and hamper well-being of children.

Other than the cultural beliefs, the community or society and even neighbourhood has a key role in maintaining the mental health of students (Herrenkohl, Sousa, Tajima, Herrenkohl and Moylan, 2008). Child abuse and neglect including sexual abuse is highly prevalent in India. These scars of child abuse may tamper a child completely to the extent of breaking their mental health to depression or other feelings of inadequacy (Deb, 2011). Street children and orphans are also at serious risk of stress and illness. Violence and exposure of students to violence can have serious impact; it has been seen that children exposed to violence tend to experience stress that affects their life in all spheres. These children who are exposed to violence and trauma like war and natural calamities like earth quakes, tsunami or other, tend to experience greater stress and later develop post-traumatic stress disorder (Wolfe, Scott, Wekerle and Pittman, 2001).

Resilience and Protective Factors

Resilience refers to an end-point identification of a child having overcome early risk exposure (that is, achieved positive outcomes or avoided negative outcomes). Protective factors are those qualities of the individual, experiences and aspects of a child's social environment that increase the likelihood of resilience on the part of those exposed to earlier risks (Herrenkohl *et al.*, 2008). Protective factors are those variables that buffer against the effects of risk factors. Resiliency is the ability to thrive in the face of adverse circumstances. Resilience seems to be particularly relevant for adolescents' life satisfaction due to the influence that academic success, social competence and avoidance of risky behaviour may exert in fostering desirable and satisfactory courses of life (Abolghasemi and Varaniyab, 2010). Understanding specific mechanisms that promote resilience in children and adolescents is an important research agenda. Risk and protective factors can be broadly grouped into four domains–child, family, school and community factors.

In general, research has found that multiple risk and protective factors are often associated with a single outcome. For example, the likelihood of stress among children is increased by temperament of the child, family, school and community risk factors, such as disturbed child-rearing, peer rejection and impoverished family conditions. Similarly, the likelihood of stress among children is decreased by the temperament of child, family, school and community protective factors such as child's social skills, positive parent–child relationship, positive peer modelling and positive social norms in the community (Masten, 2004). Many outcomes are also influenced by the same risk and protective factors. For example, neglectful and disturbed parenting can be a risk factor for stress, depression, school failure, poor physical health, physical abuse and drug use. Conversely, a positive parent–child relationship protects against all of these outcomes. The way, risk and protective factors interact to produce positive or negative outcomes at different stages of a child's development is complex and not always clearly understood. It may be that some protective factors only operate when certain risk factors are present. While risk and protective factors are common to certain outcomes, the pattern of risk and protection will vary widely from child to child. It is important to recognize the limitations of research in this area. Risk and protective factors are often only correlated with certain outcomes; they are not causally related to these outcomes. It may be that another variable better explains the relationship between the risk/protective factor and the outcome.

It was found that the top five coping strategies, frequently used by the students were religion, active coping, positive re-interpretation, planning and use of instrumental support; however, the main coping strategy was religion. These strategies are positive coping strategies which have been reported in studies as very adaptive and hasten the recovery from distress (Shahmohammadi, 2011).

Prevention

Literature shows the negative effects of stress on future stakeholders of a country. School and academic environment can be the most appropriate sites for intervention, not only because of the accessibility to the population, but because stressors reported to be experienced most frequently by the adolescents are those relating both directly and indirectly to the school environment (de Anda *et al.*, 2000). Professionals, in particular, psychologists, social workers, teachers and school administrators, need to recognize that a sizable portion of the adolescent population deals with stress in their daily life. Stress management interventions need to be coupled with attempts at reducing the number, intensity and impact of stressors in the adolescents' environment.

Conclusion and Recommendations

To be a child is to be playful and fun-loving, but in today's world, with the pressure imposed by the academic overload, expectations and other causes, children have lost their smiles. The academic overload in school and unrealistic parental demand and social expectations are imposing severe stress on students, especially among adolescent students (Shahmohammadi, 2011). Our increasing knowledge about the importance and impact of stress on young children should be put to good use in reducing stress factors for young children and in assisting children to increase coping strategies and healthy responses to the unavoidable stresses in their lives. In order to address stress and other allied problems experienced by students, there is a need to bring a national policy for recruitment of proportionate number of student counsellors in every educational institute for addressing mental health needs of the students; organize life skills training programmes for students to enhance their resilience capacity; organize special training programmes for students as to how to study systematically and prepare for the examinations; organize periodic workshops, special lecture and seminars in educational institutions for parents and teachers to sensitize them with various issues which cause stress to students.

References

Abolghasemi, A., and Varaniyab, S. T. (2010). Resilience and Perceived Stress: Predictors of Life Satisfaction in the Students of Success and Failure. *Procedia—Social and Behavioral Sciences*, 5, 748–52.

Abraham, R. R., Zulkifli, E. M. B., Fan, E. S. Z., Xin, G. N., and Lim, J. T. G. (2009). A Report on Stress among First Year Students in an Indian Medical School. *South East Asian Journal of Medical Education*, *3*(2), 78–81.

Al-Sowygh, Z. H., Alfadley, A. A., Al-Saif, M. I., and Al-Wadei, S. H., (2013). Perceived Causes of Stress among Saudi Dental Students. *King Saud University Journal of Dental Sciences*, *4*(1), 7–15.

Anders, K. (2011). *Stress and Family Relationships among College Students*. Poster Presented at the 39th Annual Western Pennsylvania Undergraduate Psychology Conference, New Wilmington, PA.

Bagana, E., Raciu, A., and Lupu, L. (2011). Self Esteem, Optimism and Exams' Anxiety among High School Students. *Procedia—Social and Behavioral Sciences* 30, 1331–38.

Barnett, R. C., Gareis, K. C., James, J. B., and Steele, J. (2003). Planning Ahead: College Seniors' Concerns About Career–marriage Conflict. *Journal of Vocational Behavior*, *62*(2), 305–19.

Bauman, S., Toomey, R. B., and Walker, J. L. (2013). Associations among Bullying, Cyber Bullying, and Suicide in High School Students. *Journal of Adolescence, 36*(2), 341–50.

Bodas, J., and Ollendick, T. H. (2005). Test Anxiety: a Cross-cultural Perspective. *Clinical Child and Family Psychology Review, 8*(1), 65–88.

Briggs–Gowan, M. J., Carter, A. S., Skuban, E. M., and McCue H. S. (2001). Prevalence of Social-emotional and Behavioral Problems in a Community Sample of 1- and 2-year-old Children. *Journal of the American Academy of Child and Adolescent Psychiatry, 40*(7), 811–19.

Byrne, D. G., Davenport, S. C., and Mazanov, J. (2007). Profiles of Adolescent Stress: The Development of the Adolescent Stres Questionnaire (ASQ). *Journal of Adolescence, 30*(3), 393–416.

Child Welfare Information Gateway (2009). *Parental Substance Use and the Child Welfare System.* Bulletins for Professionals, U.S. Department of Health and Human Services. Retrieved from: https://www.childwelfare.gov/pubs/factsheets/parentalsubabuse.cfm

Cortina, M. A., Fazel, M., Hlungwani, T. M., Kahn, K., Tollman, S., Cortina-Borja, M., and Stein, A. (2013). Childhood Psychological Problems in School Settings in Rural Southern Africa. *PLoS ONE 8(6):* e65041. Retrieved from: http://www.plosone.org/article/info%3Adoi%2F10.1371%2Fjournal.pone.0065041

Cyr, M., Wright, J., McDuff, P., and Perron, A. (2002). Intrafamilial Sexual Abuse: Brother–Sister Incest Does Not Differ from Father–Daughter and Stepfather–Stepdaughter Incest. *Child Abuse and Neglect, 26*(9), 957–73.

de Anda, D., Baroni, S., Boskin, L., Buchwald, L., Morgan, J., Ow, J., Gold, J. S., and Weiss, R. (2000). Stress, Stressors and Coping among High School Students. *Children and Youth Services Review, 22*(6), 441–63.

Deb, S. (2011). Exploitation and Harassment of Migrant Women and Girl Children Working as Domestic Assistant. In S. Sarkar and M. Srivastava (Eds.). *Globalization and Gender* (pp. 200–26). New Delhi: Rawat Publications.

Deb, S., and Modak, S. (2010). Prevalence of Violence against Children in Families in Tripura and Its Relationship with Socio-economic, Cultural and Other Factors. *Journal of Injury and Violence Research, 2*(1), 5–18.

Deb, S., and Walsh, K. (2011). Impact of Physical, Psychological, and Sexual Violence on Social Adjustment of School Children in India. *School Psychology International, 33*(4), 391–415.

Deb, S., and Walsh, K. (2010). Anxiety among High School Students in India: Comparisons Across Gender, School Type, Social Strata and Perceptions of Quality Time with Parents. *Australian Journal of Educational and Developmental Psychology, 10*(1), *18–31.*

Deb, S., Kumar, A., Bhattacharyya, B., and Jiandong, S. (2012). Parents' Perception about Children's Academic Stress and Child Care Related Issues. *Indian Journal of Health and Well-being, 3*(3), 608–12.

Demir, M. (2008). Sweetheart, You Really Make Me Happy: Romantic Relationship Quality and Personality as Predictors of Happiness Among Emerging Adults. *Journal of Happiness Studies, 9*(2), 257–77.

Demirtas, S. C., and Tezer, E. (2012). Romantic Relationship Satisfaction, Commitment to Career Choices and Subjective Well-being. *Procedia—Social and Behavioral Sciences,* 46, 2542–49.

Donaldson, S. I., Gooler, L. E., and Scriven, M. (2002). Strategies for Managing Evaluation Anxiety: Toward a Psychology of Program Evaluation. *American Journal of Evaluation, 23*(3), 261–73.

Edwardraj, S., Mumtaj, K., Prasad, J. H., Kuruvilla, A., and Jacob, K. S. (2010). Perceptions About Intellectual Disability: A Qualitative Study From Vellore, South India. *Journal of Intellectual Disability Research, 54*(8), 736–48.

Gupta, A. (2007). *Female Foeticide in India*. UNICEF—India. Retrieved from: http://www.unicef.org/india/media_3285.htm.

Hardy, M. S. (2001). Physical Aggression and Sexual Behavior Among Siblings: A Retrospective Study. *Journal of Family Violence, 16*(3), 255–68.

Herrenkohl, T. I., Sousa, C., Tajima, E. A., Herrenkohl, R. C., and Moylan, C. A. (2008). Intersection of Child Abuse and Children's Exposure to Domestic Violence. *Trauma, Violence, and Abuse, 9*(2) 84–99.

Howard, M. S., and Medway, F. J. (2004). Adolescents' Attachment and Coping with Stress. *Psychology in the Schools, 41*(3), 391–404.

Hughes, J. N., Zhang, D., and Hill, C. R. (2006). Peer Assessments of Normative and Individual Teacher–Student Support Predict Social Acceptance and Engagement Among Low-achieving Children. *Journal of School Psychology, 43*(6), 447–63.

Jewett, J., and Peterson, K. (2002). Stress and Young Children. *ERIC Digest: Clearinghouse on Elementary and Early Childhood Education,* EDO-PS-02-20. Retrieved from: http://ecap.crc.illinois.edu/eecearchive/digests/2002/jewett02.pdf.

Jewett, J., and Peterson, K. (2010). *Stress in Children*. Educational Resource Information Center (U.S. Department of Education). Retrieved from: http://www.education.com/reference/article/Ref_Stress_Young/

Jindal–Snape, D., and Miller, D. J. (2008). Challenge of Living? Understanding the Psychosocial Processes of the Child during Primary-secondary Transition through Resilience and Self esteem Theories. *Educational Psychology Review, 20*(3), 217–36.

Jobe–Shields, L., Alderfer, M. A., Barrera, M., Vannatta, K., Currier, J. M., and Phipps, S. (2009). Parental Depression and Family Environment Predict Distress in Children Prior to Stem-Cell Transplantation. *Journal of Developmental Behavioural Paediatrics, 30*(2), 140–46

Khan, R., and Cooke, D. J. (2008). Risk Factors for Severe Inter-Sibling Violence: A Preliminary Study of a Youth Forensic Sample. *Journal of Interpersonal Violence, 23*(11), 1513–30.

Kim, Y. S., Leventhal, B. L., Koh, Y. J., and Boyce, W. T. (2009). Bullying Increased Suicide Risk: Prospective Study of Korean Adolescents. *Archives of Suicide Research, 13*(1), 15–30.

Kshirsagar, V. Y., Agarwal, R., and Bavdekar, S. B. (2007). Bullying in Schools: Prevalence and Short-term Impact. *Indian Paediatrics, 44*(1), 25–28.

Kumar, A., Gireesan, A., and Deb, S. (In Press). Students' Perception of Abuse and Neglect. *Indian Journal of Health and Well-being*.

Lavigne, J. V., Gibbons, R. D., Christoffel, K. K., Arend, R., Rosenbaum, D., Binns, H., Dawson, N., Sobel, H., and Isaacs, C. (1996). Prevalence Rates and Correlates of Psychiatric Disorders Among Preschool Children. *Journal of the American Academy of Child and Adolescent Psychiatry, 35*(2), 204–14.

Linares, L. O. (2006). An Understudied form of Intra-family Violence: Sibling-to-Sibling Aggression Among Foster Children. *Aggression and Violent Behavior, 11*(1), 95–109.

Luby, J. L., Heffelfinger, A. K., Mrakotsky, C., Hessler, M. J., Brown, K. M., and Hildebrand, T. (2002). Preschool Major Depressive Disorder: Preliminary Validation for Developmentally Modified *DSM-IV* Criteria. *Journal of the American Academy of Child and Adolescent Psychiatry, 41*(8) 928–37.

Luby, J. L., Heffelfinger, A. K., Koenig–McNaught, A. L., Brown, K., and Spitznagel, E. L. (2004). The Preschool Feelings Checklist: A Brief and Sensitive Screening Measure for Depression in Young Children. *Journal of the American Academy of Child and Adolescent Psychiatry, 43*(6), 708–17.

Luby, J. L., Heffelfinger, A. K., Mrakotsky, C., Brown, K. M., Hessler, M. J., Wallis, J. M., and Spitznagel, E. L. (2003). The Clinical Picture of Depression in Preschool Children. *Journal of the American Academy of Child and Adolescent Psychiatry, 42*(3), 340–48.

Luthar, S. S., Sawyer, J. A., and Brown, P. J. (2006). Conceptual Issues in Studies of Resilience: Past, Present, and Future Research. *Annals of the New York Academy of Science, 1094*, 105–15.

Mackay, R. (2005). The Impact of Family Structure and Family Change on Child Outcomes: A Personal Reading of the Research Literature. *Social Policy Journal of New Zealand, 24*, 111–33.

Malhotra, A., Warner, A., McGonagle, A., and Lee-Rife, S. (2011). *Solutions to End Child Marriage: What the Evidence Shows*. Washington, DC: International Centre for Research on Women.

Masten, A. S. (2004). Regulatorory Process, Risk and Resilience in Adolescent Development. *Annals New York Academy of Sciences, 1021*, 310–19.

Mathers, C., and Loncar, D. (2006). Projections of Global Mortality and Burden of Disease from 2002 to 2030. *PLoS Med, 3*(11), e442.

McNamara, S. (2000). *Stress in Young People: What's New and What to Do*. New York: Continuum.

Ministry of Women and Child Development (2007). *Study on Child Abuse: India 2007*. New Delhi: Government of India.

Muhisina, S. (2013). *Students' Views on Corporal Punishment*. (Unpublished Masters Dissertation). Department of Applied Psychology, Pondicherry University. Pondicherry.

National Crime Records Bureau. (2010). *Accidental Deaths and Suicides in India: 2010*. New Delhi: Ministry of Home Affairs, Government of India.

Negovan, V., and Bagana, E. (2011). A Comparison of Relationship Between Self-esteem and Vulnerability to Depression among High School and Freshmen University Students. *Procedia—Social and Behavioral Sciences, 30,* 1324–30.

Newberry, M., and Davis, H. A. (2008). The Role of Elementary Teachers' Conceptions of Closeness to Students on Their Differential Behaviour in the Classroom. *Teaching and Teacher Education, 24*(8), 1965–85.

Pickhardt, C. (2009). *Remarriage with Adolescent: the Perils of Step Relationship*. Retrieved from: http://www.psychologytoday.com/blog/surviving-yourchildsadolescence/200909/remarriage-adolescents-the-perils-step-relationship

Raguram, R., Weiss, M. G., Keval, H., and Channabasavanna, S. M. (2001). Cultural Dimensions of Clinical Depression in Bangalore, India. *Anthropology and Medicine, 8*(1), 31–46.

Raj, L. (2011). Understanding Corporate Punishment in India. *Career Educator: An Interdisciplinary Education Journal, 1*(1), 3–18.

Sava, F. A. (2002). Causes and Effects of Teacher Conflict-inducing Attitudes towards Pupils: A Path Analysis Model. *Teaching and Teacher Education, 18*(7), 1007–21.

Sawyer, S. M., Afifi, R. A., Bearinger, L. H., Blakemore, S. J., and Dick, B. (2012). Adolescence: A Foundation for Future Health. *The Lancet, 379*(9826), 1630–40.

Schmeelk–Cone, K. H., and Zimmerman, M. A. (2003). A Longitudinal Analysis of Stress in African American Youth: Predictors and Outcomes of Stress Trajectories. *Journal of Youth and Adolescence, 32*(6), 419–30.

Selye H. (1956). *The Stress of Life.* New York: McGraw-Hill.

Shahmohammadi, N. (2011). Students' Coping with Stress at High School Level Particularly at 11th and 12th Grade. *Procedia—Social and Behavioral Sciences, 30*, 395–401.

Singh, S., Manjula, M., and Philip, M. (2012). Suicidal Risk and Childhood Adversity: A Study of Indian College Students. *Asian Journal of Psychiatry*, *5*(2), 154–59.

Sinha, R., and Jastreboff, A. M. (2013). Stress as a Common Risk Factor for Obesity and Addiction. *Biological Psychiatry*, *73*(9), 827–35.

Suldo, S. M., Shaunessy, E., and Hardesty, R. (2008). Relationships among Stress, Coping, and Mental Health in High-achieving High School Students. *Psychology in the Schools, 45*(4), 273–89.

Sundblad, G. B., Jansson, A., Saartok, T., Renström, P., and Engström, L. (2008). Self-rated Pain and Perceived Health in Relation to Stress and Physical Activity among School-Students: A 3-year Follow-up. *Pain, 136*(3), 239–49.

Tagay, Ö., and Karataş, Z. (2012). An Investigation of Attachment Styles of College Students. *Procedia—Social and Behavioral Sciences, 47*, 745–50.

Trifoni, A., and Shahini, M. (2011). How Does Exam Anxiety Affect the Performance of University Students? *Mediterranean Journal of Social Sciences*, *2*(2), 93–100.

Troop–Gordon, W., and Gerardy, H. (2012). Parents' Beliefs about Peer Victimization and Children's Socio-emotional Development. *Journal of Applied Developmental Psychology, 33*(1), 40–52.

Uitto, M. (2012). Behind Every Profession is a Person': Students' Written Memories of their Own Teacher-Student Relationships. *Teaching and Teacher Education, 28*(2), 293–301.

Verma, S., Sharma, D., and Larson, R. W. (2002). School Stress in India: Effects on Time and Daily Emotions. *International Journal of Behavioral Development*, *26*(6), 500–08.

Wang, Y., and Ollendick, T. H. (2001). A Cross-Cultural and Developmental Analysis of Self-esteem in Chinese and Western Children. *Clinical Child and Family Psychology Review*, *4*(3), 253–71.

William, J. H., and Veeh, C. A. (2012). Continued Knowledge Development for Understanding Bullying and School Victimization. *Journal of Adolescent Health, 51(1)*, 3–5.

Wolfe, D. A., Scott, K., Wekerle, C., and Pittman, A. L. (2001). Child Maltreatment: Risk of Adjustment Problems and Dating Violence in Adolescence. *Journal of the American Academy of Child and Adolescent Psychiatry*, *40*(3), 282–89.

CHAPTER 13

Applied Positive Psychology: Enabling Communities to Thrive

Nov Rattan Sharma

Maharishi Dayanand University, Rohtak, Haryana

Abstract Psychology is the study of individual and group behaviour patterns. There are large numbers of recognized divisions which focus on the subject matter of psychology, specifically in a particularistic manner. However, psychology has shown satisfactory progress in almost every domain of life, but the current emphasis of psychology is on positive psychology which has emerged as a very significant discipline of knowledge. It has successfully endorsed its attendance in almost every sphere of science. Positive psychology has been accepted as an authentic and effective commander of living well at every stage of life. It seems that positive psychology has the potential to lead the entire humanity in the direction of a meaningful life. In fact, tenets of positive psychology seem to be the closest to the basic essence of Indian philosophy. Therefore, one may find the roots of modern positive psychology in the ancient Indian philosophy. There are sufficient empirical evidences which demonstrate that pillar concepts of positive psychology are significantly contributing to the development of virtues and strengths in the individual as well as communities. The scientific body of knowledge credited to positive psychology further reinforces social and community psychologists to apply principles of positive psychology in different forms of communities for enabling them to thrive and flourish rapidly as well as successfully. Sizable researches here in India and abroad have been initiated in the recent past to carry out these applications at individual, group, family, school, society and work organization levels. The concept such as positive schooling, positive parenting, positive psychological capital and positive business organizations are in the testing stage around the globe. Research indices yielded so far clearly indicate that people in general and communities in particular may be empowered in such a way that maximum realization on their part can be achieved. This chapter will critically examine the associations and impact of positive psychology-based interventions on the functioning of communities.

In most Indian and foreign universities psychology was part and parcel of the Philosophy Department, and it was the hard work and visionary wisdom of some philosophers who advocated that psychology has the potential to be developed as an independent discipline of knowledge. But the basic question was to determine the nature, scope and methodology of psychology. There is an evolutionary history of psychology, where you may find its roots, however, in the present chapter, the author has no intention to elaborate on the long past of psychology but at the same time he has all intentions to mention that due to its early affiliation to philosophy, the nature, scope and methodology of psychology are bound to be affected by its mother discipline. The definition of psychology itself crossed different viewpoints before it reached the state of general acceptance. Readers are aware that in the beginning psychology was considered as the study of soul/mind/consciousness, etc. However, the scientific bent of the then philosophers' mind was not satisfied. Rather, the viewpoints to explain psychology as study of soul/mind/consciousness were rejected on one or two basic grounds. Psychologists failed to demonstrate these concepts and more so there was no scientific methodology to measure them. However, for some time such concepts floated in the field without proper investigation. Many schools at different locations across the world were also trying to define psychology from their own point of view. Structuralism, Functionalism, Behaviourism, Psychoanalysis and Gestalt being the major schools of psychology forwarded their considered opinion and put forward the same along with the scope and methodology of psychology. The three primary questions—(i) what is psychology? (ii) what is the subject matter of psychology? and, (iii) what is the most suitable methodology of psychology?—raised at that time, were very simple in nature but at the same time the most difficult to answer amicably. The efforts and words invested in to answer these questions were assembled, critically evaluated and discussed at various platforms across the globe. Psychologists swiftly acted to sort out differences and pick up the similarities. The final picture about the nature of psychology, which emerged largely, was the study of human behaviour.

However, psychology at its first stage was neutral, neither positive nor negative. Scope of psychology was around the behaviour, that is, normal and abnormal behaviour, behaviour across different age groups and study of those biopsychosocial factors that affect the behaviour of the organism. The proposed methodology for the study of behaviour had to be scientific and therefore, observation method was considered to be the most suitable to study behaviour.

Positive Psychology

In view of its neutrality, psychology travelled a long journey where role of various schools of psychology also became non-existent. In the beginning, research works undertaken at different laboratories was also of a very primary kind, however, it gathered momentum to create a variety of researches from the pure to the applied kind. Large number of divisions also came into existence to take care of diversified fields of psychology. The expansion also proved helpful in the identification of currented psycho-social and educational needs of mankind. Very soon, psychologists started realizing that unless they converted their pure researches into the applications of psychology, particularly for the welfare of the organism, the real objective of psychology as a science would remain unachieved. This thinking resulted in the development of many more applied branches of psychology and 'positive psychology' is one among this series. The American Psychologist devoted its millennial issue to the field of positive psychology with a clear indication on the three goals—(i) study of positive emotions; (ii) study of positive character; and (iii) study of positive institutions. Seligman and Csikszentmihaly (2000) highlighted the need and importance of the emergence of this science of positive psychology. This shifting was in contrast to the earlier approach of psychology where focus was on the negative side of life using concepts like depression, anxiety and trauma. However, the paradigm shift from the negative to the positive side of life is not very old but has resulted in revolutionary changes in the mindset of modern researchers of social sciences in general and of psychology in particular. The positive psychology perspective is wholesome for making good life possible by fulfilling the lives of common people, helping them in preventing illness and also enhancing psychological as well as overall health. The recognition of positive emotions in life further created a research domain where current researches have demonstrated that positive emotions may be used as an energy bank to combat the prevailing negative feelings. Positive emotions like love, affection, kindness, etc., are like major human life values. It is well said that there is no life without human values. More so, love is the under current of almost every positive emotion and human value that give power to each value to function and grow.

Similarly, positive character is another milestone concept of positive psychology recognized and put for testing in laboratory. However, talking about the positive character was the focus of traditional moral philosophy and religion but contemporary positive psychology has also entered in its scientific assessment as well as promotion. The values in action (VIA), classification of that strength that make the good life possible are the major objectives of many research projects around the world. As per Value in Action Institute (VIA), there are six core virtues, namely, wisdom and knowledge; courage; humanity; justice; temperance and transcendence and 24 positive traits/strengths under these six virtues. However, a final word on it is expected after exhaustive empirical validation (Peterson and Seligman, 2004.) The third pillar of positive psychology is the study of positive institutions. The significance of institution, let it be family or school or an organization or a small or large workplace, has been reconstructed with the concept of positive institutions. It provides ample importance to those issues and concepts which facilitate democratic and mutual strong bindings amongst its members which further help in supporting the first two concerns of positive psychology. The scope of positive psychology is increasing day by day and more and more research is being taken up on its core concepts. The best part with positive psychology is its methodology. It is just like a scientific revolution and sizable empirical data has been gathered with scientific procedures to validate and revalidate the significance of the tenets of positive psychology.

Applied Positive Psychology

Psychology of today is positive. Majority of its research projects directly or indirectly show some associations with positive psychology. However, it is also true that positive psychology of today seems to be strongly correlated with the age-old concepts of Indian philosophy. That is why we can see the linkage amongst traditional moral philosophy, religious acts and the principles of positive psychology. There is an academic as well as professional pressure on almost every branch of knowledge to have a strong applied connection. In order to ease out the pressure, there is a trend of upcoming courses in the fields such as Applied Physics, Applied Mathematics and Applied Psychology. In psychology itself, the prevalence of courses like applied social psychology and applied positive psychology is not uncommon. The emphasis on such courses is much higher than their respective pure courses because of the need of the hour. For example, scope of positive psychology, that is, its pure branch has included the new concepts such as resilience, flow, happiness, gratitude, forgiveness and life satisfaction for the understanding of their meaning, nature and study. In applied positive psychology, these concepts are not only studied and assessed but their applications are also adjudged. There are researches which are monitoring these applications in different areas particularly in some restricted areas for example, health, well-being, making life good, that too at individual as well as group levels. It is beyond the scope of the present chapter to enumerate these issues; however, author is making an effort to assimilate these issues under the following three major sectors: (a) living (b) schooling and (c) working.

Since birth to death, development is an integral part of the human life. Man, being a social organism, affects others and gets affected by others. Personality make up of a child also comes across different developmental stages and family traditions and ritual systems provide adequate background for his/her socialization. Formation of self image and identity also get shaped during adolescence. Most of the time an adolescent is either at home with parents and siblings or at school with teachers and peer group members, therefore, both the places, that is, home and school are the most crucial determinants of behaviour domain of this age group. However, home and school factors have been the significant subject matter of research but emergence of positive psychology has changed the impetus of recent investigations. Positive parenting and positive schooling are the current outcome of applied positive psychology, where social scientists have successfully demonstrated skill development training programme pertaining to home and school targeting desirable results. Same is true for an adult, when he/she approaches the workplace. The physical structure of an organization, interpersonal relationship among co-workers and communication between employee and employer directly influence the wellness of the entire unit. The concept like positive psychological capital is worth mentioning to highlight the importance of workplace. In other words, enabling these three units of communities that is, home, school and organization will go a long way to justify the objectives of applied positive psychology.

Self and Communities

An individual is the smallest building unit of a group/family. Family is the foundation stone of a society, rather, individuals living together make a family and families living together make a community. It means that individual, family and community are inseparable. More so, family system of behaviour constitutes the individual system of personality, likewise community system determines the family system. However, a sick system may lead to faulty interactions amongst members resulting in stress as well as mental health problems, whereas, a healthy system will create and facilitate an atmosphere of wellness and facilitate a feel of good life for each other. Unfortunately, most of the times families and communities are recognized for narrow gains such as solemnizing marriage, showing voting behavior and meeting favourable means. The above restricted meanings of the communities are not to the tune of applied positive psychology. In view of the psychosomatic nature of many diseases, the underlying role of families and communities need to be maximized. The interdependence and interplay between organism and organizations including family and community do not allow the researcher to compartmentalize these three identified sectors (living, schooling and working) into distinct independent divisions. With due recognition of this fact, author is making an effort to enlist the need to enable the community at large.

It has been widely recognized that the basic nature of human being is not only to help others but also to extend the self to others for collaboration in working to design as well as achieve common, larger and even super-ordinate goals including having faith and trust in members of the group. In view of the transpersonal (beyond self) urge of humans, community group approach get its expected impetus. Not only this, by enabling the self and then institutions like family, school and organizations, all their members may also approach towards success and accomplishment in almost every field of life. However, the impact of community approach has been widely and broadly noticed in the area of health. The high medical cost and non-availability of sufficient medical and para-medical staff in health sector are another two compelling limiting factors for a country like India. Therefore, India and all other nations need to empower their citizens and institutions to achieve the goal of 'health for all'. In psychology, the development of the branch on community health psychology is the most fitted remedy for many kinds of broad spectrum illnesses because there is an agreement now that the most humane and economical way to maintain, guard and even enhance the overall health and well-being of an individual, family and large groups is only through prevention. These days, health professionals have also realized that positive thoughts may also act as preventive vaccine. The purpose of individualist and collective approach remains to sensitize, build up and apply preventive measures. The community level of analysis and collective action in identifying the health barriers and psycho-social mediation in respect to local circumstances were the key issues in determining the potential role of partnership in creating healthy organizations.

The community approach may be further given a shape of effective tool while creating and implementing social policies. Tripathi and Sinha (2013) have observed that psychologists have been given a very little role in framing policies particularly in the domains of education and public health at macro as well as micro levels. They have further emphasized that psychological knowledge may play a vital role in the process of decision making leading to national development. The research contributions by social scientists at national and international levels have significantly highlighted that small steps may result in big outcomes if implemented strategically by enabling people and institutions in a single go. Therefore, the three units of the stakeholders are to be taken in mixed order for empirical verification and validation. Some of the researches supervised and undertaken by the author are summarized in this chapter.

Hooda and Sharma (2007) recognized that recently health has secured more optimistic and positive definition and it can be described along a continuum with illness and wellness. In view of the multidimensional state of health, researchers advocate holistic approach to health with a harmonious balance among its different domains. However, there are some psychological factors, for example, intelligence which may influence holistic health. Some studies were conducted where different domains of positive health were systematically investigated in relation to general intelligence, emotional intelligence (Hooda, Sharma and Yadava, 2008, 2011a) and social intelligence (Hooda, Sharma and Yadava, 2009, 2013). On the basis of the investigation conducted on three hundred working adults, it was observed that there is a significant negative association between general intelligence and positive health indices. In other words, very high level of intelligence is not necessary for good positive health. On the other hand, emotional intelligence and social intelligence were the key associates of positive health. General health, happiness, satisfaction with life, spirituality and life orientation were found to be associated at significant levels with the sub scales of emotional intelligence and social intelligence (Hooda, Sharma and Yadava, 2011b). Most remarkably, appraisal of negative as well as positive emotions, the indicators of emotional intelligence, was observed as the significant predictor of positive health. Interpersonal skills and flexibility, another component of emotional intelligence, was positively associated with happiness and life satisfaction. Patience also contributed towards the prediction of positive health. In case of social intelligence, the factors which significantly accounted for positive health to a large extent are sense of humour and memory. It was observed that the ability to create and enjoy humour not only helps in the development of positive health but also creates a feel good atmosphere around oneself to maintain it even in adverse circumstances. The overall findings support the view that irrespective of age, gender and marital status it is rather more important to be emotionally as well as socially connected to self and others (Hooda, Sharma, and Yadava, 2010, 2011c). It further strengthens that enabling the individuals and their group/community at emotional and social constructs creates an atmosphere of cooperation and care which may further develop the feeling of wellness.

Tuyet and Sharma (2008) conducted a study where 100 participants were selected from *Vipassana* meditation (VM) training groups. VM is one of the most ancient Indian techniques of meditation. Buddha discovered this method and put it in practice as a remedy for the universal ills. During the period of Buddha, large number of people in India were seeking its training and also got freedom from the bounds of sufferings by its practice. Even the literature of that time hinted that people practicing *Vipassana* used to attain high levels of achievements in almost every sphere of life. The participants of VM groups were compared with the non-*vipassana* meditation group. More so, this comparison was mostly focused on health parameters and before and after procedure was used to check the health status at two occasions. Mainly four domains of health, that are physical (heart rate and blood pressure), psychological (happiness, subjective well-being, anger expression), social and spiritual, were put for investigation before *Vipassana* and after the *Vipassana* meditation training camp. It was evident from the findings of the study that blood pressure (diastolic and systolic) and heart rate of the participant before and after VM occasions were statistically different. It also means that VM significantly reduced the heart rate and blood pressure levels of the participants to normal. In case of psychological health, VM greatly enhanced happiness and well-being status, and reduced negative feelings and anger levels. At the same time social and spiritual component of health also got noticeable enhancement after attending the camp of *Vipassana* meditation. In sum, all the different domains of health are linked and VM affects almost every domain. As reported by the investigator, practice of *Vipassana* relaxes the body, makes the mind happy and peaceful, creates sociability and generates the meaning in life. The result of this overall harmonious effect is not only the optimal functioning of bodily organs but also change in attitude of the participants towards self and others. The change towards positivism further increases the value of peace and harmony in life which undoubtedly contribute to social and community development. The findings recommend that courses/camps on *Vipassana* meditation may be organized regularly at learning, living and working places. Such programmes may also be legitimately conducted in tune with other community health awareness camps to strengthen preventive module of health.

Modern life style violated many basic principles of health and people pay little attention on what and when to eat; where and how one lives and how one thinks. As per our old system of medicine *(Ayurveda)*, there are four fundamental principles of life style—*ahara* (food), *vihar* (recreation), *achara* (routine) and *vichara* (thinking). In order to stay healthy, these four principals should be given proper attention. Even modern medicine system recognizes the importance of life style. A healthy life style decreases the chances of falling sick where as unhealthy life style not only aggravates diseases like hypertension, diabetes, coronary heart disease (CHD), cancer and asthma but also creates conditions prone to general illness. Although, there is a cluster of life style diseases, this concept in itself is rather diffused. Since life style included the way people live and reflect social values, attitudes and cultural patterns of socialization, they are learnt through interactions between self and significant others (Menon and Kennedy, 2004).

Sangwan and Sharma (2009) conducted a study to investigate the role of lifestyle among hypertensive, diabetics, asthmatics and normal people. A sample of 400 subjects participated in the study with 100 respondents in each of the four groups. The findings of the study successfully demonstrated the role of life style in all three disease groups; life style of the hypertensive was found to be the worst. It implies that unhealthy life style with faulty behavioural patterns and life long personal habits may lead to hypertension. However, life style indicators of diabetics and asthmatics were also not healthy in comparison to the normals but hypertensive individuals remained the most vulnerable culprits of unhealthy life styles. As reported earlier, life style is a learnt phenomena and a resultant effect of social and cultural activities. It implies that health education disseminated in small or large family/community groups on life style diseases will not only create a dent on the spread of these diseases but also help in the development of healthy life style among new generation.

Another major bottleneck in healthy and happy communities is the aggression among adolescents. You may find aged people blaming young generation on account of lack of tolerance and patience, increased level of anger and hostility and immediate gratification of needs. The clash between young adolescents and their parents or grand parents is not uncommon. There is a felt need to train the youngsters and teenagers for their unnecessary and unhealthy out bursts in their living, schooling and playing premises. Pankaj and Sharma (2012) made sincere efforts in the investigation on the assessment and management of adolescents' anger. However, anger is a natural emotion expressed by every human but its expression may take various forms, for example, sober, mild or even a full blown rage. It may trigger many physiological, behavioural and cognitive effects. Therefore, assessment and management of adolescents' anger is the primary duty of social scientists. Even when anger is a natural emotion with functional survival value, there are right and wrong ways to deal with the angry moments and it can become the source of many kinds of physical, mental, emotional, social and even legal problems if not managed adequately. Five hundred adolescents studying in different public schools of the national capital region were put for the assessment of their anger. In the first phase, only assessment of anger was done; and, in the second phase of study, only those students participated whose anger levels were found to be above the third quartile. The investigator designed some anger management strategies and the same were used either in single or in the combination of two. The obtained results indicated that anger if managed satisfactorily may even enhance subjective well-being of the students and the enhanced well-being status may further promote their scholastic and personal achievements. The results have demonstrated a strong linkage between anger expression and other health parameters, where anger control builds up healthiness and expression causes detrimental effects on health. Researchers compared the three (psychological, yogic and physiological) intervention procedures of anger management. However, all the three intervention programmes not only significantly reduced the anger level of the high anger groups but also psychological as well as yogic procedures yielded maximum benefits (Pankaj and Sharma, 2012). It implies that by creating psycho-educative technique and yoga, anger and its bad effects may be restrained. Managing anger may also act as a preventive medicine in communities and in this way self and institutions may be empowered.

Cardiovascular diseases (CVD) are the major public health concern being the number one cause of death worldwide. It has become the largest killer in developed countries, also rapidly becoming one in developing countries. India, like many other developing countries, is facing an enormous health concern. Once upon a time, curative health care used to be the major focus of doctors and health professionals. Individuals too were only concerned about their health when something went wrong. Rightly so, the world has increased its focus on prevention and relapse prevention. However, some risk factors for heart disease cannot be altered (for example, family history, gender and age), but the majority are modifiable conditions that may respond to intervention (for example, cigarette smoking, hypertension, physical inactivity, diabetes, obesity, lack of social support, depression, anxiety, stress, Type-A behaviour pattern). Furthermore, we know why and how CHD develops and, more importantly, how to prevent it. We must take full advantage of this to promote a comparatively healthy life for CHD patients and others prone to this illness. As health research continues to discover more pathways to disease prevention, it becomes clearer that medical intervention alone cannot accomplish all the work ahead. It is important to involve psychological and social intervention for achieving desirable outcomes. Recent studies have also demonstrated that meditation-based stress reduction programmes are equally useful interventions for patients with various medical and psychological symptoms (Nehra and Sharma, 2010).

The aim of a very recent study conducted by Nehra, Sharma, Kumar and Nehra (2014) was to examine the expected beneficial effects of a mindfulness-based stress-reduction (MBSR) programme on the psycho-physiological parameters in CHD patient. Findings of the study have shown that MBSR had the hypothesized effects in the sample of distressed individuals from the CHD population. Results supported previous findings regarding the effectiveness of MBSR on decrease of depression, anxiety, stress, health complaints and Type-A behaviour pattern. For the fact, accepting without judgement and observing mainly sensory phenomena may correlate with improvement on psychological measures. A potential explanation of these findings may be that the focus of the intervention is on enhancing mindfulness in general, which leads to accepting one's feelings and thoughts, and observing bodily and physical sensations. However, MBSR group did not show any improvement on the physiological measures. The analysis of data of the study indicates that there has been a remarkable decline in psychological parameters after the intervention. The sample that was subjected to psychological interventions reported a significant alleviation in their symptoms. Respondents further reported that their physical well-being had improved as a result of a relief from their symptoms. They could also resume their normal activities with much vigour. Counselling of the significant others had helped in changing their attitude towards the client and reinforcing them intermittently. The findings conclude that incorporating a psychological intervention, in addition to drug therapy, into a cardiac rehabilitation programme effectively modified cardiac risk factors and may improve recovery and prognosis. The implications of this investigation include following the principle that awareness is bliss and it is the duty of family and other social organizations to create it. Such kinds of psycho-educative programmes at the institutional level help us to harness maximum benefits to the greatest extent in communities.

Health professionals would agree that health involves much more than simply absence of disease. It is quite possible for a person to be free of disease but still not enjoy a vigorous, satisfying life. Health does not exist in isolation and may be influenced by a complex of environmental, social, economical, cultural, religious and personal factors ultimately related to each other (Nandal and Sharma, 2011). Religious factors like rituals, religiosity, spirituality, and religious beliefs are inseparable from an individual's psyche. In general sense, rituals related to birth, death, marriage, initiation, healing, harvest or religious observance are found in all known cultures, and appear to have been performed for thousands of years. They speak to people's core emotions and reveal values that a society holds dearest. Ritual is neither the recapitulation of the profane nor the imprisonment of the sacred, rather the enactment of a referring-to, orienting-toward, focusing-upon and opening-to the transcendent sacred. It is a unified system of beliefs and practices relative to sacred things viewed as multidimensional in nature. Through most of the history of modern psychology, this term has been both, an individual and an institutional construct.

Religion, according to the *Gita*, is that which is inherent in the soul. It cannot be changed, and it is universally the same for all living entities. External faiths are material reflections of the inherent spiritual quality of the soul. On the other hand, spirituality is 'transpersonal awareness arising spontaneously or through meditative and other practices beyond ordinary explanations, and to which are attributed an inspiring and guiding meaningfulness, often attributed to a Deity' (Benor, 2006, p. 467). A study conducted on a sample of 400 educated adults demonstrated that spirituality came out to be a significant contributing factor on life satisfaction and holistic health in the total group. In case of subjective well-being, the significant contributing factors came out to be the rituals followed by religiosity (Nandal, Sharma and Yadava, 2013).

There are factors such as happiness, well-being, satisfaction, health and positive communities and institutions that foster good lives and gainful employment along with the economic and environmental policies that promote harmony and sustainability. Practical applications of positive psychological capital include helping employees and organizations to identify their strengths and use them to increase and sustain their respective levels of wellness. As India is one of the populated places on earth, human capital plays a very important role in fuelling the Indian economy. Therefore, the newly emerging concept of positive psychological capital would be particularly relevant when Indian organizations are preparing to compete in the global economy. Researches on the positive psychological capital states (Hope, Optimism, Resiliency and Efficacy) and employees' wellness have been almost ignored in the field of organizational psychology. For further extension in the positive organization behaviour for organizations, Luthans, Luthans and Luthans (2004) provide a combined construct, that is, psychological capital as a main contributor to the competitive advantage of organizations. They refer to the positive psychological resources which include confidence, hope, optimism and resilience. Psychological capital consists of positive psychological capabilities that are assessable, open to enlarge and manageable. As positive psychological capital can be expanded or developed there are many interventions to build up the psychological capital states in employees. Mindfulness is one of those proposed but not studied much by empirical research. Mindfulness is currently defined in psychological terms as characterized by paying total attention to the present moment with a non-judgemental awareness of the inner and outer experience (Marlatt and Kristeller, 1999). Chauhan and Sharma (2013) investigated the relationship between employees' wellness and positive psychological states. This study also examines the effect of meditation on a target group having low score on positive psychological capital. Results have been discussed in the light of correlation and multiple regression analyses. The study has observed that understanding, assessment and management of positive psychological capital is very important. Findings reveal that wellness is positively and significantly correlated with four states of positive psychological capital (efficacy, hope, optimism and resilience) of employees. Efficacy and hope are significant contributors on overall wellness, job satisfaction, subjective well-being and gainful employment. It means that employees with high self efficacy and hope have more wellness feeling revealed by high level of job satisfaction, subjective well-being and gainful employment. Optimism and resilience significantly contributed to overall wellness, level of job satisfaction, subjective well-being and gainful employment. It also means that employees who are sturdy or having power of resilience had the tendency of high subjective well-being. Results consolidated the view that mindfulness meditation was effective positive psychological capital intervention, which increases the positive psychological capital of employees. Findings further demonstrated that the effectiveness of the intervention programme was maintainable up to a longer period. Positive psychological capital states—efficacy, hope, resilience and optimism of the employees are also affected by the intervention programme. As far as the efficacy of this intervention is concerned, the findings of present investigation have favoured the significant positive effect of intervention on positive psychological capital (Chauhan and Sharma, 2013). The implication of the study may be seen mainly in organizational psychology, human resource management, workplace intervention programmes, for therapists and trainers, organizations and positive psychology particularly with the objective to enable and empower the employees and their institutions.

The Indian view of health takes into consideration the achievement and maintenance of balance between all aspects of life such as social, emotional and spiritual. The focus is not only on curing the disease but also promoting positive experiences to health and dealing with the life conditions. To a large extent, factors such as where one lives, the state of one's environment, genetics, one's income and education level, and one's relationships with friends and family all have considerable impact on health and well-being, whereas the more commonly considered factors such as psycho-social factors, participation in civic activities and social engagement, strong social networks, feeling of trust, feeling of power and control over life decisions, supportive family structure, positive self-esteem, healthy conditions and environments, effective health services, healthy life styles and use of health care services often have lesser impact. Thus, the context of people's lives determines their health, and so blaming individuals for having poor health or crediting them for good health is inappropriate. Individuals are unlikely to be able to directly control many of the determinants of health such as—accidents and injuries, diet, nutrition and health, immunization, tobacco smoking, physical activity, alcohol, obesity, drug misuse, etc. After the positive psychology revolution, every concept seems to feel connected with its principles. So is the case of health and well-being in present study. It has been noticed that positive capacities do have potential to affect health and well-being of an individual significantly (Mamta and Sharma, 2013a).

Each and everyone have a capacity which may be stable and help to live better life. Even the knowledge of these capacities and awareness that the same are present within the self make a difference in making and creating change in attitude towards life events. In the study by Mamta and Sharma (2013b), positive capacities were divided in three parts, namely, present, past and future orientation positive capacities. Humour, meaning in life and resilience were taken as present orientation positive capacities. Forgiveness, gratitude and life satisfaction were taken as past orientation positive capacities. Hope, optimism and self-efficacy were taken as future orientation positive capacities. The primary aim of the study was to find out positive capacities as correlates of health and well-being among working and retired persons. Most of the present orientation positive capacities (humour, meaning in life and resilience) were found significantly correlated with life stress, coping skills, mental health, and response to illness, health habits, adult health history and health care utilization amongst retired persons while significantly correlated with social resources and health beliefs and attitudes in working persons. Whereas the past orientation positive capacities (forgiveness, gratitude and life satisfaction) were found significantly correlated with life stress, social resources and mental health among working persons but it was also found that most of past orientation positive capacities were significantly associated with coping skills, response to illness, health habits and health beliefs and attitudes in retired persons while equally significantly correlated with health care utilization in working and retired persons. Lastly, future orientation positive capacities (hope, optimism and self-efficacy) were found significantly correlated with life stress, social resources, mental health, response to illness, health care utilization and health beliefs and attitudes in working persons while it was also found that future orientation positive capacities were significantly correlated with coping skills and health habits in retired persons. However, it was found that most of the future orientation positive capacities equally significantly associated with adult health history in working and retired persons. Thus, most of the positive capacities, that is, present, past and future orientations were found significantly correlated with well-being manifestation among retired persons (Mamta and Sharma, 2013b).

Conclusion

In conclusion, the overall research outputs shown in preceding sections clearly indicate that applied positive psychology has wider scope and application in producing desirable results. The respective findings are not only applicable at individual level but also have strong tendency to affect group functioning. Be it the case of positive capacities, emotional and social intelligence, positive psychological capital, ritual, religiosity and spirituality, management of anger, building up healthy life style, practicing some meditational procedure or dealing with chronic diseases like CHD, diabetics, Asthma research demonstrates that enabling the self will help develop feeling of having control over his/her health related concerns. This feeling of control is further translated into action at living (home), studying (school) and working (organization) places. The effects are not limited to the individual; they are bound to roll towards the family and communities. People with healthy mind and body will create healthy family and happy communities. However, focus at multiple domains may be suffering of an individual but letting others not suffer is also of paramount importance. How can you achieve the goal of enabling the organizations (home, school and industry) without serving their smallest units adequately? Applied positive psychology with its concepts and applications has provided sample methodological, scientific, workable and effectively functional provisions which one may apply at every small or big step of life without much problem. One has to get up and start working on them. Think small but think for all. It is better to try and fail than fail to try.

References

Benor, D. J. (2006). *Personal Spirituality: Science, Spirit, and the Eternal Soul*. Medford, NJ: Wholistic Healing.

Chauhan, K., and Sharma, N. R. (2013). *Positive Psychological Capital and Employees' Wellness*. Unpublished Doctoral Thesis. Maharshi Dayanand Unversity, Rothak.

Hooda, D. and Sharma, N. R. (2007). *Intelligence as a Correlate of Positive Health*. Unpublished Doctoral Thesis. Maharshi Dayanand Unversity, Rothak.

Hooda, D., Sharma, N. R., and Yadava, A. (2008). Emotional Intelligence as a Predictor of Positive Health in Working Adults. *Journal of Indian Health Psychology, 2*(2), 196–207.

Hooda, D., Sharma, N. R., and Yadava, A. (2009). Social Intelligence as a Predictor of Positive Psychological Health. *Journal of Indian Academy of Applied Psychology, 35*(1), 143–50.

Hooda, D., Sharma, N. R., and Yadava, A. (2010). Intelligence as Contributor of Positive Health Across Gender. *Indian Social and Psychological Studies*, *3*(2), 59–70.

Hooda, D., Sharma, N. R., and Yadava, A. (2011a). Level of Emotional Intelligence and Positive Health. *Psyber News—An International Magazine of Psychological Research, 2*(4), 147–54.

Hooda, D., Sharma, N. R., and Yadava, A. (2011b). Emotional Intelligence and Spiritual Health Among Adults. *Journal of Indian Academy of Applied Psychology, 37*(2), 246–50.

Hooda, D., Sharma, N. R., and Yadava, A. (2011c). Effect of Age, Gender and Marital Status on Positive Subjective Experiences. *Indian Journal of Psychological Science, 2*(1), 1–9.

Hooda, D., Sharma, N. R., and Yadava, A. (2013). Effect of Social Intelligence on Positive Living. *Journal of Positive Psychology, 2*(1), 16–23.

Luthans, F., Luthans, K., and Luthans, B. (2004). Positive Psychological Capital: Going Beyond Human and Social Capital. *Business Horizons, 47*(1), 45–50.

Mamta and Sharma, N. R. (2013a). Resilience and Self-efficacy as Correlates of Well-being Among the Elderly Persons. *Journal of the Indian Academy of Applied Psychology, 39*(2), 281–88.

Mamta and Sharma, N. R. (2013b). *Positive Capacities as Correlate of Health and Well-being: A Comparative Study of Working and Retired Persons*. Unpublished Doctoral Thesis. Maharshi Dayanand Unversity, Rothak.

Marlatt, G. A., and Kristeller, J. L. (1999). Mindfulness and Meditation. In W. R. Miller (Ed.), *Integrating Spirituality into Treatment: Resources for Practitioners* (pp. 67–84). Washington, DC: American Psychological Association.

Menon, S. D., and Kennedy, A. (2004). *Health Psychology and Stress Management*. Hyderabad: Apollo Hospitals Educational and Research Foundation.

Nandal, T., and Sharma, N. R. (2011). *Rituals, Religiosity and Spirituality as Correlates of Health*. Unpublished Doctoral Thesis. Maharshi Dayanand Unversity, Rothak.

Nandal, T., Sharma, N. R., and Yadava, A. (2013). A Correlation Study of Religiosity and Health. *Journal of Indian Health Psychology*, *8*(1), 75–87.

Nehra, D. K., and Sharma, N. R. (2010). *Comparative Efficacy of CBT and MBSR Program on Psycho-Physiological Parameters in CHD Patients.* Unpublished Doctoral Thesis. Maharshi Dayanand Unversity, Rothak.

Nehra, D. K., Sharma, N. R., Kumar, P., and Nehra, S. (2014). Efficacy of Mindfulness-based Stress Reduction Progarmme in Reducing Perceived Stress and Health Complaints in Patients with Coronary Heart Disease. *Dysphrenia,* 5(1). 19–25.

Pankaj and Sharma, N. R. (2012). *Assessment and Management of Adolescent Anger.* Unpublished Doctoral Thesis. Maharshi Dayanand Unversity, Rothak.

Peterson, C., and Seligman, M. E. P. (2004). *Character Strength and Virtues: A Hand Book and Classification.* Oxford, England: Oxford University Press.

Sangwan, S., and Sharma, N. R. (2009). *Type a Behavior Pattern and Lifestyle Amongst Hypertensives, Diabetics and Asthmatics: A Comparative Study.* Unpublished Doctoral Thesis. Maharshi Dayanand Unversity, Rothak.

Seligman M. E. P., and Csikszentmihalyi, M. (2000). Positive Psychology: An Introduction. *American Psychologist,* 55(1), 5–14.

Tripathi, R. C., and Sinha, Y. (2013). *Psychology, Development and Social Policy in India.* New Delhi: Springer.

Tuyet, N. T., and Sharma, N. R. (2008). *Effect of Vipassana Meditation on Health.* Unpublished Doctoral Thesis. Maharshi Dayanand Unversity, Rothak.

CHAPTER 14

Positive Psychology in Suicide Prevention

Archana, Vijay Parkash and Updesh Kumar

Defence Institute of Psychological Research, Delhi

Abstract Suicide is an extreme form of behaviour that occurs due to an individual's inability to manage stress. It develops through an interplay of psychological, situational, cognitive and affective factors that place individuals at risk. It is due to these various factors that it becomes difficult to predict and prevent suicide. There has been enormous increase in the past decade in research pertaining to positive psychology that focuses on enhancing human potential. Positive psychology serves as a buffer against the risk of suicide. It is concerned with the well-being and optimal functioning of an individual that leads to greater personal fulfilment and positive living. Hope, resilience, optimism, autonomy, positive mood, sense of worth, life satisfaction and character strengths are considered vital constructs in the area of positive psychology that has gained immense amount of empirical support in promoting adaptive functioning following extreme stress. In this chapter we attempt to understand the role of positive psychology in preventing suicide and how positive living helps in leading calm, serene and stress-free life.

We live in a tough and challenging world. In the present global economy, change is inevitable, thus creating ambiguities and uncertainties in situations that are stress inducing. With increasing complexity in our life style, the level of stress has been rising at a phenomenal rate. Although everyone experiences stress due to adverse life situations, the intensity of it varies depending upon the context and the capability of the person to cope with it. Most people are able to handle these tough times and make best out of it. But at times, the inability to adapt to such challenges may force an individual to take his life. Under such circumstances, it becomes essential to develop skills required for adjusting to difficult conditions, thus having a stress-free and positive life. The reflections of positive life are seen in positive psychology that focuses on promoting mental health and well-being. Good mental health and well-being are essential elements for functioning well and having a good quality of life. It strengthens and prepares an individual for coping with the challenges of life.

During the past few years, positive psychology has gained tremendous impulsion. Positive psychology is 'nothing more than the scientific study of ordinary human strengths and virtues, one that revisits the average person' (Sheldon and King, 2001, p. 216.) In today's complex world, it has become essential to focus on the essence of positive psychology that facilitates positive traits, nurtures strengths and cultivates positive social relationships. Positive psychology fosters hardiness and self-enhancement. Hardiness relates to finding essence in life by believing that one has the potential to grow from both positive and negative life events (Kobasa, Maddi and Kahn, 1982). Research reveals that hardiness reduces the effects of stress through positive coping and social support (Florian, Mikulincer and Taubman, 1995). Self-enhancement relates to the tendency of favouring one's own self by reducing one's negative traits (Taylor and Brown, 1988). It is found to be associated with better adaptation and growth during challenging life events (Bonanno, Rennicke and Dekel, 2005). Hardiness and self-enhancement are considered as effective psychological tools that are essential for sustaining optimal mental health and overcoming the vulnerabilities of life. The principles of positive psychology adopted for effective coping during stress usually focus on self-esteem, positive emotions, building trust, considering oneself as a responsible citizen and contributing effectively towards the welfare of the society.

Ironically, research has given much importance to negative traits as compared to positive traits. The emerging trends in positive psychology bring a more pragmatic and rational view of human nature that focuses on the strengths and capabilities of the individual without overlooking their flaws and limitations. Promoting positive psychology indicates a shift from the deficit approach to building an adaptive approach that helps to raise above all the complexities of life. It shows that our perception, belief and attitude towards life play a significant role in contributing towards well-being, happiness and quality of life.

The role of positive psychology in suicide prevention is highly significant to understand, since it provides a wide array of constructs (like hope, optimism, positive emotions, strengths, virtues, resilience and positive future orientation) that helps an individual towards adapting positively to the adverse conditions of life. These constructs are often used as an intervention strategy for uplifting the psychological, emotional and mental health. They are modifiable and are often capable of reducing depressive symptoms (Seligman, Steen, Park and Peterson, 2005).

Positive psychology is a move towards a strength-based approach that plays a vital role in improving the mental health and well-being of individuals, thus reducing the risk of self-defeating behaviour. Poor mental, physical and social health leads to suicidal tendencies (Juurlink, Herrmann, Kopp and Redelmeier, 2004). In a study carried out by Druss and Pincus (2000), it was observed that significant relationship exists between suicidality and poor physical health as well as suicidality and poor social health. By enhancing psychological skills, the vulnerabilities to suicide can be countered, thus making life worth living for individuals and preparing them to successfully adapt to the challenges of life.

Understanding Suicide

Suicide is a growing concern today. It is a complex act that occurs across the world, irrespective of culture, race and religion. The reasons that people choose to take their own life are very complex. Suicide stems from the thought to end one's life as a solution to the problems in one's life. People commit suicide when their vital needs are not adequately met and they remain uncertain about their life. This indecisiveness often leads them to no other option than to engage in self-defeating behaviour. The survival of an individual depends upon the way he perceives the whole situation. Each individual performs multiple roles in his life and the imbalance in these roles often leads to stress. The life event may be same, but its impact may vary from person to person depending upon the situation as well as the skills and capabilities of an individual to handle stress under adverse conditions. The failure in coping with traumatic life events often results in vulnerability that predisposes an individual to respond in a negative way to difficult circumstances. Life events are crucial in deciding the fate of an individual. A negative life event can trigger suicide-related behaviour in some people.

An individual's ability to cope with the pressures of adverse life events depends on many factors that include his genes, family history, psychiatric illnesses, perceived failures, mental health, well-being, quality of life, interpersonal relationships and social support systems. Self-defeating behaviour is often preceded by stressful life events that may be either acute or chronic in nature. Acute life events include strained family relations, bereavement, legal problems (Brent *et al.*, 1993) and chronic life events are associated with chronic pain and physical illness (Braden and Sullivan, 2008). Suicide is a multi-faceted problem and as the problems multiply within the individual, the act of suicide becomes much more intense and impulsive.

It is extremely difficult to determine the instant and enduring factors that are responsible for being suicidal. The instant factors are the ones that often act as precipitants to suicide, such as severe depression, loss of job or a loved one and drug or alcohol intoxication. And the enduring factors have a cumulative effect and often stem from childhood neglect and trauma, disturbed home environment, family history of suicide, impulsivity and a lack of interpersonal support and so on. Since it's an impulsive act, people vary in their expression regarding the degree of indication for such behaviour. For example, some people give prior warning before taking their own lives and some end their life without any warning.

In order to understand the concept of suicide, it is essential to know the cultural context in which an individual resides, works and carries out his routine activities. The problem of suicide needs to be addressed at the stage of ideation itself, so that it can be avoided at an appropriate time. For preventing suicide, one needs to lay emphasis on embracing personal strengths and virtues. The approach of positive psychology towards suicide prevention is considered as a significant means for adapting positively to stressful events.

Interplay of Risk and Protective Factors in Suicide

The role of positive psychology in suicide prevention depends upon the constellations of risk and protective factors. The inter-relationship of both risk and protective factors play a vital role in affecting one's behaviour for committing suicide. Risk factors increase the likelihood of suicidal thoughts and behaviour and protective factors eliminate these negative thoughts of killing oneself by increasing one's capability to deal with challenging situations. Risk and protective factors are on two extreme ends, for example, on a component of social support, both these factors are on the opposite directions wherein risk factors are indicated through social isolation and protective factors are reflected through social connectedness. Likewise, on the dimensions of self-esteem and self-appraisal, these factors again are on the opposite ends, wherein high self-esteem and positive self-appraisal strengthens the association between protection and suicide (Johnson, Gooding, Wood and Tarrier, 2010).

Prevalence of Risk Factors

The risk factors for suicide involve a complex interplay of psychological, environmental, social, genetic and neurobiological factors, which differ according to what part of the suicidal process they predict. The prevalence of all these factors predisposes an individual to behave in a specific manner that varies from situation to situation or changes as per the circumstances. It also depends upon one's capability to grow and learn from challenges of life. Under such circumstances the main challenge is to identify those adaptive strategies that help in coping with stressful life events.

Some of the factors responsible for increasing the likelihood of suicide include impulsiveness (Fawcett, 2001; Zouk, Tousignant, Seguin, Lesage and Turecki, 2006), aggressiveness (Conner, Swogger, and Houston, 2009), feeling of hopelessness (Brezo, Paris, and Turecki, 2006), difficulty in decision-making and problem solving, (Keilp, Gorlyn, Oquendo, Burke, and Mann, 2008) and mental disorder (Kessler, Borges, and Walters, 1999). Mood, alcohol/substance abuse, impulse-control, psychotic and personality disorders are also found to be strongly related to high suicidal rate (Hawton, Houston, Haw, Townsend and Harris, 2003). Other numerous researches reveal a negative relationship between suicide and life satisfaction as well as suicide and feelings of happiness (Voracek, 2009). Study findings indicate that people high on depression and hopelessness are found to have diminished quality of life and high risk for suicide (De Man, 1999). Lack of social support, negative self-concept and pessimism leads to depression, which increases the likelihood of suicidal behaviour (Au, Lau and Lee, 2009).

Joiner (2005) proposed that there are three variables that indicate an individual's risk for committing suicide. These are: perceived burdensomeness, thwarted belongingness and acquired capability for suicide. Thwarted belongingness refers to the lack of maintaining meaningful relationships with others. Perceived burdensomeness is a feeling of considering oneself as a burden on others and failing in one's effort in making a significant contribution towards society. Acquired capability for suicide indicates an individual's capability to enact a lethal suicide attempt. Certain personality traits like introversion, irritation and neuroticism are related to high risk for suicide, whereas other factors like conscientiousness and extraversion buffer suicide (Brezo, Paris and Turecki, 2006).

Though it is not possible to prevent all the risk factors for suicide, certain principles of positive psychology may be instilled in an individual that may help in preventing suicide. Restoring the constructs of positive psychology, impede the complex pathways to suicidality.

So far most of the research pertaining to suicide has focused on understanding the risk factors (Prinstein, Boergers, Spirito, Little and Grapentine, 2000; Jackson and Nuttall, 2001). However, there has been lack of empirical support on the presence of protective factors towards suicide prevention. Recently, the focus has been shifted towards identifying the protective factors that mitigate the risk for suicide (World Health Organization, 2002).

Prevalence of Protective Factors

Before strengthening protective factors for suicidal behaviour, an individual needs to have a thorough understanding of the risk factors. The accumulation of many risk factors makes an individual more vulnerable to commit suicide (Durlak, 1998). Researchers have found that the presence of protective factors serves as a buffering effect for suicidal prevention. During stress, the presence of certain traits promote adaptive functioning. For example, positive reinterpretation, religiousness and acceptance coping eliminate the effects of suicidal behaviour (Park, Cohen and Murch, 1996). Also individuals high on emotional intelligence are more likely to adapt well under stressful situations (Cha and Nock, 2009). Social support appears to be one of the most tested factors that protects against suicidal behaviour (Qin and Mortensen, 2003).

For countering the effects of risk, one needs to develop strategies that foster resilience and instill protective factors. Also, by enhancing life skills and promoting mental health and well-being, protective factors can be restored. By having awareness regarding mental illness and by developing community networks, the problem of suicide can be prevented to some extent. Protective factors, thus, can be enhanced through building self-esteem, personal competence and fostering positive relationships.

Principles of Positive Psychology in Suicide Prevention

Although difficult, suicidal behaviour can be avoided in any setting. Suicide prevention strategies should address issues pertaining to positive mental health, well-being and social connectedness. Friends and family members are the greatest source of support during the time of crisis. They are the best indicators for identifying the rate of suicidal risk. Therefore, efforts need to be put in for strengthening the protective factors and fostering social networks for mitigating the stressors. Cultivating the approach of positive psychology may serve beneficial to protect an individual from committing the act of suicide. Given below are some of the constructs of positive psychology that help in preventing suicide.

Hope

An individual encounters stress in every sphere of life that hinders his goal-oriented behaviour. The circumstances may be stressful, but the hope an individual carries builds a positive attitude in dealing with the adversities. Hope can be classified as true hope and false hope (Groopman, 2004). False hope fails to identify the risk and true hope acknowledges it. According to Gillham (2000), hope is defined as a challenging force that drives an individual to cope with the psychological problems, thus leading towards the goal. The concept of hope involves two components: affective and cognitive. Affective components focus on positive emotions that help in forming flourishing relationships with others and cognitive component emphasizes on thought processes that create pathways towards attaining goals (ibid.).

Hopelessness occurs due to the interplay of stressful events along with an individual's awareness about the available means for overcoming the barriers. Those who have doubt on their capabilities often perceive the situation as highly emotion provoking and are more likely to give up and feel hopeless about the situation. And those who feel confident about their abilities to overcome difficult situations, often embrace a feeling of hopefulness.

Hope has emerged as a significant predictor of health and well-being (Siegal, 1986). Hope is 'a positive motivational state that is based on an interactively derived sense of successful (a) agency, and (b) pathways (planning to meet goals)' (Snyder, Irving and Anderson, 1991, p. 287). Agency is the willpower of an individual that provides the motivation and strength of mind for achieving goals. It is a sustained effort towards attaining the objectives, irrespective of the hurdles that come in between. On the other hand, pathways are the waypower that involves having belief in oneself in accomplishing goals. Further, it also suggests that irrespective of the obstructions on the way of reaching a target, the individual has enough capability to produce alternative routes for attaining means. People who are hopeful are high on optimism, problem solving skills and motivation (Snyder, 2000). Also, these people are more skillful in looking for alternative substitutes for achieving their targets. Hopeful people do not ponder on the negative issues rather they focus on the constructive ways for solving their problems. Hopelessness appears to be the key determinant and the most reliable predictor for suicidal thought and behaviour.

The principles of positive psychology believe in eliminating the feelings of hopelessness and strengthening hopefulness. It believes in looking forward to positive thoughts and overcoming the feelings of negativity. On one hand, hopefulness is considered as an important element for suicide prevention (Beck, Brown, Berchick, Stewart and Steer, 1990) and on the other, hopelessness has emerged as one of the most significant predictors of suicidal behaviour (Hawton and van Heeringen, 2009; McMillan, Gilbody, Beresford and Neilly, 2007). In a research carried out by Beck, Weissman, Lester and Trexler (1974), it was found that the feeling of hopelessness results in depression and suicidal behaviour. A study by Ran *et al.*, (2005) revealed hopelessness as the strongest predictor of past suicide attempts. The negative impact of hopelessness can be reduced by the presence of several factors like having a strong social support, being resilient, having a positive orientation towards life and by fostering the feelings of hope.

Social Support

Social support serves as a protective factor for both well-being and suicidal thoughts (Conner, Duberstein, Conwell, Seidlitz and Caine, 2001; Duberstein, Conwell, Conner, Eberly, Evinger and Caine, 2004). Social support sought through the strength and bonding of relationship with family and community is found to be related to lower rates of suicide and increased life satisfaction. Lack of social support observed in the form of lack of acceptance and isolation, are found to be related to high suicidal rates (Spruijt and de Goede, 1997). Individuals who lack in family bonding and have less social networks are more prone to suicidal behaviour. People having close supportive relationships benefit good health and happiness as compared to those who lack in establishing social network with others (Ryff and Singer, 2000). Other studies confirm that strong social relationships with friends and family members often result in less sickness and longer lives; whereas lack of social network leads to unhappiness, loneliness and depression (Berscheid, 2003). Family cohesion in the form of mutual involvement and strong emotional support buffers against suicidal behaviour. Perceived social support from others buffers the negative effects of stress and contributes to good health and well-being. It acts as a safeguard for suicide. And, the most influential support comes from family that attenuates against the risk factors associated with suicide. Research reveals social connection and social support as the most significant protective factors against suicide (Conner *et al.*, 2001).

Failure in one's effort to fulfil the need for affiliation may often result in deficiency of having social connection with others, thus leading to suicidal behaviour. Thwarted belongingness has emerged as one of the most significant risk factors for committing suicide (Joiner, 2005).

Optimism

Optimism signifies a positive approach with a determination to go on in life. It plays a vital role in mental and physical health. Scheier and Carver (1985) define optimism-pessimism in terms of outcome expectancies, where optimism relates to an outlook of anticipating positive outcomes and pessimism relates to looking forward to negative outcomes. An individual's belief and attitude depends upon the way the situation is perceived. Optimistic individuals have a positive approach towards life and these approaches to optimism are of two types: dispositional optimism and explanatory style. Dispositional optimism focuses on seeing the brighter side of the life and believing the future to be positive (Scheier and Carver, 1992). According to these researchers, optimistic people are self-regulated and display more confidence in overcoming the challenges of life, whereas pessimists often doubt their capabilities and express lack of confidence in dealing with the life stressors. Research shows that optimism is positively associated with several traits like: self-esteem, happiness, good mental and physical health and well-being (Carver and Scheier, 2002). As compared to optimists, pessimists are more likely to suffer from depression and anxiety. Explanatory style suggests an individual's way of defining events that are unconstructive in nature. Optimistic individuals will always see the positive side of stressful situations and pessimists will always find stress as permanent aspect of life and dubious to change the future. Optimism is associated with better life satisfaction, lower levels of distress and less psychological maladjustment (Chang and Sanna, 2001).

Optimism serves as a foundation for motivation. Optimists perceive negative events as temporary in nature that can be handled effectively if the person displays strong self-confidence.

They are highly flexible and are more likely to adopt active problem-solving strategies while dealing with the life events. Optimists usually do not disengage themselves from attaining a goal and thereby a negative relationship emerges between optimism and suicidal ideation (Hirsch, Conner and Duberstein, 2007).

Optimism as one of the important constructs of positive psychology, serves as a protective factor against suicidal behaviour. According to Gillham (2000), optimism is an expectation on the part of the individual that life will bring good results. Individuals who are optimistic are more likely to overcome stressors in their lives. Optimism is one of the most dynamic predictors of behaviour that helps people in attaining their goals by overcoming the challenges of life. During stressful conditions, an optimistic individual involves in focused coping strategies that help them in adapting to the negative life events. In contrast to this, pessimists are more likely to engage in self-defeating behaviour when dealing with the negativities of life (Carver and Scheier, 2002).

Optimism and hopelessness are negatively related to each other, where optimism counters the feelings of hopelessness and acts as a buffer against suicide (O'Connor and Cassidy, 2007). As compared to optimists, pessimists often exhibit their incapability in attaining a goal, which leads to anguish and frustration. They are also likely to disengage themselves from the routine activities of life, thereby increasing the probability of suicide (Scheier and Carver, 1987). Optimists barely withdraw themselves from those situations that block their goal-oriented behaviour and therefore are less likely to commit suicide (Scheier, Weintraub and Carver, 1986). Abundant studies have been carried out in this area that show negative relationship between optimism and suicide, suggesting that the effects of suicide can be mitigated by promoting positive outcome expectancies. People high on optimism are likely to have strong support system, elevated self-confidence and reduced feeling of committing suicide (Hirsch, Conner and Duberstein, 2007). These factors contribute in enhancing one's health and well-being (Heisel and Flett, 2004). Optimism along with perceived social support fosters strong bond and social network with others. It provides a protection against distress, thus mitigating the suicidal risk.

Positive Future Orientation

Future orientation is an approach that focuses on future events and outcomes of one's acts. People with positive future orientation have focused goals and feel confident in using adaptive strategies for achieving their goals. According to Hirsch *et al.* (2006), future orientation refers to the skills of an individual that helps them in developing an optimistic approach towards life and identifying means for attaining set targets. Future orientation is a cognitive component and serves as a significant predictor of optimism that stresses on remaining hopeful in future. People who view their future in positive direction are less likely to commit suicide. Szanto and colleagues (Szanto *et al.*, 2001) suggest that the risk of suicide can be eliminated by altering negative thoughts and by holding a positive approach about future. Therefore, by transforming future orientation, the risk for suicide can be lessened. People who are optimistic about their future are less prone to self-defeating behaviour as compared to people who have pessimistic approach and express high feelings of hopelessness (Hirsch, Wolford, LaLonde, Brunk and Morris, 2007). Positive future orientation is associated with well-being, healthy psychological adaptation, low levels of depression and decreased risk for suicide (Ormel *et al.*, 1997).

Positive Emotions

Emotions come in the form of positive and negative affect that are complementary to each other. They form a strong base for people's emotional lives. According to broaden-and-build theory developed by Fredrickson (2001), positive emotions broaden one's thinking ability by constructing resources that are beneficial for enhancing the mental health and well-being. Positive emotions foster positive thinking, personal growth and strong human relationships resulting in healthy and successful living. Taken together, it expands one's outlook, counteracts negative emotions, cultivates optimism, enhances resilience and improves one's emotional state that is essential in dealing with all the challenges of life. Negative emotions can lead to variety of dysfunctions resulting in anxiety disorders (Ohman, 1993), depression (Nolen-Hoeksema, Morrow and Fredrickson, 1993) and suicide (Chen and Dilsaver, 1996). Positive emotions result in more flexible behaviour during adversities and help in recovering from negative experiences.

Resilience

Resilience is the capacity on the part of the individual to recover from the adverse life events. It is the ability to bounce back, recover from or adjust to misfortune or change (Burns, 1994). Resilient individuals possess strengths to endure stress and adverse life conditions, while emphasizing on positive growth and forming quality relationships (Connor, 2006; Fredrickson, Tugade, Waugh and Larkin, 2003). Having positive emotions increases an individual's thought processes thereby enhancing their coping skills and resilience and also helps in recovering effectively from daily hassles of life (Fredrickson, 2001).

According to Gilligan (2000), resilience is based on three aspects: sense of secure base, self-esteem and self-efficacy. A secure and sound base grows out of social support an individual receives, having strong connectedness with others and establishing a daily routine in one's lifestyle. Self-esteem is based on maintaining a safe and cordial relation with others and attaining success in achieving one's goals. Self-efficacy relates to the ability of having self-confidence in taking one's own decision in life.

The concept of well-being and resilience are interconnected, wherein well-being facilitates resilient behaviour and resilience contributes to enhance well-being. Research findings suggest that in the organizational set up, well-being is found to be related to resilient outcomes. For example, a study by Keyes (2002) reveals that people high on good mental health are less likely to take off from work, have focused goals, are resilient and display less health-related problems. When individuals are exposed to life-threatening circumstances, those who are resilient are likely to adapt effectively under such conditions and exhibit reduced adverse mental health outcomes. The varying life experiences help an individual to learn, grow and develop intervention strategies for dealing effectively with the adversities of life. And, if the individual fails in his efforts to grow, then he is likely to become vulnerable. Vulnerability predisposes a person to respond negatively to stressful situations. For example, the support from family often results in building resilience; whereas, lack of family support makes an individual vulnerable and more prone to end one's life. Numerous researches show that the resilient individuals experience significant well-being and are less likely to get involved in self-defeating behaviour (for example, Sawyer, Guidolin, Schulz, McGinnes, Zubrick and Baghurst, 2010). An inverse association between resilience and suicide indicates that the perceptions and beliefs of an individual play a vital role in buffering the harmful impact of risk. Resilient individuals usually display positive self-appraisals in their behaviour. These positive self-appraisal thoughts reflect an individual's ability to deal with the emotions positively. During distressing situations, the positive self-appraisals buffer against suicidal thoughts and behaviour (Johnson, Gooding, Wood and Tarrier, 2010).

Concluding Remarks

Today, it has become a great challenge for preventing suicide and preserving life. Prevention in its best form can be attained through enhancing wellness at psychological, social and emotional levels. There is a need to increase efforts for identifying protective factors for suicide. These efforts will build the foundation for future research in suicide. In recent years, significant research has been carried out pertaining to the nature of suicidal behaviour, but more efforts need to be put forth for conceptualizing how the constructs of positive psychology work together in overcoming the acts of suicide. It should take into account those adaptive coping strategies that cultivate positive psychological growth, supportive interpersonal relations, meaningful living, future orientation, strong motivation and determination for achieving goals. Positive psychology plays a very important role in maintaining a healthy outlook on life. The role of positive psychology is to bring out the best qualities of an individual in the best possible way by building on his strengths and virtues that acts as a buffer against suicide, thus leading to a balanced life. Future research requires an increasing effort on the part of the researchers to have a better understanding of the role of positive psychology in suicide prevention.

References

Au, A. C. Y., Lau, S., and Lee, M. T. Y. (2009). Suicide Ideation and Depression: The Moderation Effects of Family Cohesion and Self-concept. *Adolescence, 44*(176), 851–68.

Beck, A. T., Brown, G., Berchick, R. J., Stewart, B. L., and Steer, R. A. (1990). Relationship between Hopelessness and Suicide: A Replication with Psychiatric Outpatients. *American Journal of Psychiatry, 147*(2), 190–95.

Beck, A. T., Weissman, A., Lester, D., and Trexler, L. (1974). The Measurement of Pessimism: The Hopelessness Scale. *Journal of Consulting and Clinical Psychology, 42*(6), 861–65.

Berscheid, E. (2003). The Human's Greatest Strength: Other Humans. In L. G. Aspinwall and U. M. Staudinger (Eds.), *A Psychology of Human Strengths: Fundamental Questions and Future Directions for a Positive Psychology* (pp. 37–48). Washington, DC: American Psychological Association.

Bonanno, G. A., Rennicke, C., and Dekel, S. (2005). Self-enhancement among High-exposure Survivors of the September 11th Terrorist Attack: Resilience or Social Maladjustment? *Journal of Personality and Social Psychology, 88*(6), 984–98.

Braden, J. B., and Sullivan, M. D. (2008). Suicidal Thoughts and Behavior among Adults with Self-reported Pain Conditions in the National Comorbidity Survey Replication. *Journal of Pain, 9*(12), 1106–15.

Brent, D. A., Perper, J. A., Moritz, G., Baugher, M., Roth, C., Balach, L., and Schweers, J. (1993). Stressful Life Events, Psychopathology and Adolescent Suicide: A Case Control Study. *Suicide and Life-Threatening Behavior, 23*(3), 179–87.

Brezo, J., Paris, J., and Turecki, G. (2006). Personality Traits as Correlates of Suicidal Ideation, Suicide Attempts and Suicide Completions: A Systematic Review. *Acta Psychiatrica Scandinavica, 113*(3), 180–206.

Burns, T. E. (1994). *From Risk to Resilience: A Journey with Heart for Our Children, Our Future*. Dallas, Texas: Marco Polo Publishers.

Carver, C. S., and Scheier M. F. (2002). Optimism. In C. R. Snyder and S. J. Lopez (Eds.), *Handbook of Positive Psychology* (pp. 231–43). New York, NY: Oxford University Press.

Cha, C. B., and Nock, M. K. (2009). Emotional Intelligence is a Protective Factor for Suicidal Behavior. *Journal of the American Academy of Child and Adolescent Psychiatry, 48*(4), 422–30.

Chang, E. C., and Sanna, L. J. (2001). Optimism, Pessimism and Positive and Negative Affectivity in Middle-aged Adults: A Test of a Cognitive-affective Model of Psychological Adjustment. *Psychology and Aging, 16*(3), 524–31.

Chen, Y., and Dilsaver, C. C. (1996). Lifetime Rates of Suicide Attempts among Subjects with Bipolar and Unipolar Disorders Relative to Subjects Without Axis 1 Disorders. *Biological Psychiatry, 39*(10), 896–99.

Conner, K. R., Duberstein, P. R., Conwell, Y., Seidlitz, L., and Caine, E. D. (2001). Psychological Vulnerability to Completed Suicide: A Review of Empirical Studies. *Suicide and Life-Threatening Behavior, 31*(4), 367–84.

Conner, K. R., Swogger, M. T., and Houston, R. J. (2009). A Test of the Reactive Aggression-suicidal Behavior Hypothesis: Is There a Case for Proactive Aggression? *Journal of Abnormal Psychology, 118*(1), 235–40.

Connor, K. M. (2006). Assessment of Resilience in the Aftermath of Trauma. *Journal of Clinical Psychiatry*, 67(2), 46–49.

De Man, A. F. (1999). Correlates of Suicide Ideation in High School Students: The Importance of Depression. *The Journal of Genetic Psychology, 160(1),* 105–14.

Druss, B., and Pincus, H. (2000). Suicidal Ideation and Suicide Attempts in General Medical Illnesses. *Archives of Internal Medicine, 160*(10), 1522–26.

Duberstein, P. R., Conwell, Y., Conner, K. R., Eberly, S., Evinger, J. S., and Caine, E. D. (2004). Poor Social Integration and Suicide: Fact or Artifact? A Case-controlled Study. *Psychological Medicine, 34*(7), 1331–37.

Durlak, J. (1998). Common Risk and Protective Factors in Successful Prevention Programs. *American Journal of Orthopsychiatry, 68*(4), 512–20.

Fawcett, J. (2001). Treating Impulsivity and Anxiety in the Suicidal Patient. *Annals of the New York Academy of Sciences, 932*(1), 94–102.

Florian, V., Mikulincer, M., and Taubman, O. (1995). Does Hardiness Contribute to Mental Health During a Stressful Real-life Situation? The Roles of Appraisal and Coping. *Journal of Personality and Social Psychology, 68*(4), 687–95.

Fredrickson, B. L. (2001). The Role of Positive Emotions in Positive Psychology. *American Psychologists, 56*(3), 218–26.

Fredrickson, B. L., Tugade, M. M., Waugh, C. E., and Larkin, G. R. (2003). What Good are Positive Emotions in Crises? A Prospective Study of Resilience and Emotions Following the Terrorist Attacks on the United States on September 11th, 2001. *Journal of Personality and Social Psychology*, *84*(2), 365–76.

Gillham, J. (2000). *The Science of Optimism and Hope.* Radnor, PA: Templeton Foundation Press.

Gilligan, R. (2000). Adversity, Resilience and Young People: The Protective Values of Positive School and Spare Time Experiences. *Children and Society, 14*(1), 37–47.

Groopman, J. (2004). *The Anatomy of Hope: How People Prevail in the Face of Illness*. New York: Random House.

Hawton, K., Houston, K., Haw, C., Townsend, E., and Harris, L. (2003). Comorbidity of Axis I and Axis II Disorders in Patients Who Attempted Suicide. *American Journal of Psychiatry, 160*(8), 1494–1500.

Hawton, K., and van Heeringen, K. (2009). Suicide. *The Lancet*, *373*(9672), 1372–81.

Heisel, M. J, and Flett, G. L. (2004). Purpose in Life, Satisfaction with Life and Suicide Ideation in a Clinical Sample. *Journal of Psychopathology and Behavioral Assessment, 26*(2), 127–35.

Hirsch, J. K., Duberstein, P. R., Conner, K. R., Heisel, M. J., Beckman, A., Franus, N., and Conwell, Y. (2006). Future Orientation and Suicide Ideation and Attempts in Depressed Adults Ages 50 and over. *American Journal of Geriatric Psychiatry, 14*(9), 752–57.

Hirsch, J. K., Conner, K. R., and Duberstein, P. R. (2007). Optimism and Suicide Ideation Among Young Adult College Students. *Archives of Suicide Research, 11*(2), 177–85.

Hirsch, J. K., Wolford, K., LaLonde, S. M., Brunk, L., and Morris, A. P. (2007). Dispositional Optimism as a Moderator of the Relationship Between Negative Life Events and Suicide Ideation and Attempts. *Cognitive Therapy and Research, 31*(4), 533–46.

Jackson, H., and Nuttall, R. L. (2001). Risk for Preadolescent Suicide-related Behaviours: An Ecological Model. *Child and Adolescent Social Work Journal, 18*(3), 189–203.

Johnson, J., Gooding, P., Wood, A. M., and Tarrier, N. (2010). Resilience as Positive Coping Appraisals: Testing the Schematic Appraisals Model of Suicide (SAMS). *Behaviour Research and Therapy*, *48*(3), 179–86.

Joiner, T. E. (2005). *Why People Die by Suicide*. Cambridge, MA, US: Harvard University Press.

Juurlink, D. N., Herrmann, N., Szalai, J. P., Kopp, A., and Redelmeier, D. A. (2004). Medical Illness and the Risk of Suicide in the Elderly. *Archives of Internal Medicine, 164*(11), 1179–84.

Keilp, J. G., Gorlyn, M., Oquendo, M. A., Burke, A. K., and Mann, J. J. (2008). Attention Deficit in Depressed Suicide Attempters. *Psychiatry Research, 159*(1–2), 7–17.

Kessler, R. C., Borges, G., and Walters, E. E. (1999). Prevalence of and Risk Factors for Lifetime Suicide Attempts in the National Comorbidity Survey. *Archives of General Psychiatry, 56*(7), 617–26.

Keyes, C. L. M. (2002). The Mental Health Continuum: From Languishing to Flourishing in Life. *Journal of Health and Social Behavior, 43*(2), 207–22.

Kobasa, S. C., Maddi, S. R., and Kahn, S. (1982). Hardiness and Health: A Prospective Study. *Journal of Personality and Social Psychology, 42*(1), 168–77.

McMillan, D., Gilbody, S., Beresford, E., and Neilly, L. (2007). Can we Predict Suicide and Non-fatal Self-harm with the Beck Hopelessness Scale? A Meta-analysis. *Psychological Medicine, 37*(6), 769–78.

Nolen-Hoeksema, S., Morrow, J., and Fredrickson, B. L. (1993). Response Styles and the Duration of Episodes of Depressed Mood. *Journal of Abnormal Psychology, 102*(1), 20–28.

O'Connor, R. C., and Cassidy, C. (2007). Predicting Hopelessness: The Interaction Between Optimism/Pessimism and Specific Future Expectancies. *Cognition and Emotion, 21*(3), 596–13.

Ohman, A. (1993). Fear and Anxiety as Emotional Phenomena: Clinical Phenomenology Evolutionary Perspectives, and Information-processing Mechanisms. In M. Lewis and J. M. Haviland (Eds.), *Handbook of Emotions* (pp. 511–36). New York: Guilford Press.

Ormel, J., Kempen, G. I. J. M., Penninx, B. W. J. H., Brilman, E. I., Beekman, A. T. F., and van Sonderen, E. (1997). Chronic Medical Conditions and Mental Health in Older People: Disability and Psychosocial Resources Mediate Specific Mental Health Effects. *Psychological Medicine, 27*(5), 1065–77.

Park, C. L., Cohen, L. H., and Murch, R. L. (1996). Assessment and Prediction of Stress-related Growth. *Journal of Personality, 64*(1), 71–105.

Prinstein, M. J., Boergers, J., Spirito, A., Little, T. D., and Grapentine, W. L. (2000). Peer Functioning, Family Dysfunction and Psychological Symptoms in a Risk Factors Model for Adolescent Inpatient's Suicidal Ideation Severity. *Journal of Clinical Child Psychology*, *29*(3), 392–405.

Qin, P., and Mortensen, P. B. (2003). The Impact of Parental Status on the Risk of Completed Suicide. *Archives of General Psychiatry, 60*(8), 797–802.

Ran, M. S., Xiang, M. Z., Mao, W. J., Hou, Z. J., Tang, M. N., Chen, E. Y. H., and Conwell, Y. (2005). Characteristics of Suicide Attempters and Non Attempters with Schizophrenia in a Rural Community. *Suicide and Life Threatening Behavior, 35*(6), 694–701.

Ryff, C. D., and Singer, B. (2000). Interpersonal Flourishing: A Positive Health Agenda for the New Millennium. *Personality and Social Psychology Review, 4*(1), 30–44.

Sawyer, M. G., Guidolin, M., Schulz, K. L., McGinnes, B., Zubrick, S. R., and Baghurst, P. A. (2010). The Mental Health and Well Being of Adolescents on Remand in Australia. *Australian and New Zealand Journal of Psychiatry*, *44*(6), 551–59.

Scheier, M. F., and Carver, C. S. (1985). Optimism, Coping and Health: Assessment and Implications of Generalized Outcome Expectancies. *Health Psychology, 4*(3), 219–47.

Scheier, M. F., and Carver, C. S. (1987). Dispositional Optimism and Physical Well-being: The Influence of Generalized Outcome Expectancies on Health. *Journal of Personality, 55*(2), 169–10.

Scheier, M. F., and Carver, C. S. (1992). Effects of Optimism on Psychological and Physical Well-being: Theoretical Overview and Empirical Update. *Cognitive Therapy and Research, 16*(2), 201–28.

Scheier, M. F., Weintraub, J. K., and Carver, C. S. (1986). Coping with Stress: Divergent Strategies of Optimists and Pessimists. *Journal of Personality and Social Psychology, 51*(6), 1257–64.

Seligman, M. E., Steen, T. A., Park, N., and Peterson, C. (2005). Positive Psychology Progress: Empirical Validation of Interventions. *American Psychologist, 60*(5), 410–21.

Sheldon, K. M., and King, L. (2001). Why Positive Psychology Is Necessary. *American Psychologist, 56*(3), 216–17.

Siegal, B. S. (1986). *Love, Medicine, and Miracles: Lessons Learned About Self-healing from a Surgeon's Experience with Patients.* New York: Harper and Row.

Snyder, C. R. (Ed.). (2000). *Handbook of Hope: Theory, Measures, and Applications.* San Diego: Academic Press.

Snyder, C. R., Irving, L., and Angerson, J. R. (1991). Hope and Health: Measuring the Will and the Ways. In C. R. Snyder and D. R. Forsyth (Eds.), *Handbook of Social and Clinical Psychology: The Health Perspective* (pp. 285–305). Elmsford, NY: Pergamon.

Spruijt, E., and de Goede, M. (1997). Transitions in Family Structure and Adolescent Well-being. *Adolescence, 32*(128), 897–911.

Szanto, K., Mulsant, B. H., Houck, P. R., Miller, M. D., Mazumdar, S., and Reynolds, C. (2001). Treatment Outcome in Suicidal vs. Non-suicidal Elderly Patients. *The American Journal of Geriatric Psychiatry, 9*(3), 261–68.

Taylor, S. E., and Brown, J. D. (1988). Illusion and Well-being: A Social Psychological Perspective on Mental Health. *Psychological Bulletin, 103*(2), 193–210.

Voracek, M. (2009). National Intelligence, Suicide Rate and Subjective Well-being. *Perceptual and Motor Skills, 109*(3), 718–20.

World Health Organisation (2002). *Suicide Prevention in Europe. The WHO European Monitoring Survey on National Suicide Prevention Programmes and Strategies.* Copenhagan: WHO Regional Office for Europe.

Zouk, H., Tousignant, M., Seguin, M., Lesage, A., and Turecki, G. (2006). Characterization of Impulsivity in Suicide Completers: Clinical, Behavioral and Psychosocial Dimensions. *Journal of Affective Disorders, 92*(2–3), 195–204.

INDEX

Q

R